# A NEW APPROACH TO HUMAN SOCIAL EVOLUTION

This book provides an important examination into the role of evolution of human traits of dominance as central to understanding social and political events, proposing a new view on human social evolution. It also examines basic biological universal needs and behavioural profiles of non-human living beings, from which humans share essential survival components. It invites readers to think critically about the psychological evolution of the human brain. Using comparative psychology, it argues that the core of human behaviour lies in the ancient, animal, universal set of survival resources hidden under various socialization profiles. However, it generally fails to replace drives of dominance and aggression for physical and social survival. Genuine replacement of those primal behavioural drives would require fundamental neuro-socio-behavioural changes. This book supports the thesis that without education and the promotion of universal values involving environmental protection and individual opportunities to evolve, there will be negative consequences for individuals and communities. This book represents a critical tool for students of behavioural sciences, anthropology, politics, and evolutionary neurosciences, and will also greatly benefit other readers, such as teachers and professional researchers.

**Jorge A. Colombo, MD, PhD** is a former Full Professor at the University of South Florida (USA) and Principal Investigator at the National Research Council (CONICET, Argentina). He is also a former fellow of several international organizations, including NIH (USA), von Humboldt Foundation (Germany), DAAD (Germany), and the British Royal Society.

# A NEW APPROACH TO HUMAN SOCIAL EVOLUTION

## Persistence of Ancient Drives in Behaviour and Development

*Jorge A. Colombo*

Routledge
Taylor & Francis Group
NEW YORK AND LONDON

Designed cover image: Getty Images via Maksim Tkachenko

First published 2025
by Routledge
605 Third Avenue, New York, NY 10158

and by Routledge
4 Park Square, Milton Park, Abingdon, Oxon OX14 4RN

*Routledge is an imprint of the Taylor & Francis Group, an informa business*



*British Library Cataloguing-in-Publication Data*
A catalogue record for this book is available from the British Library

ISBN: 9781032945866 (hbk)
ISBN: 9781032951188 (pbk)
ISBN: 9781003571544 (ebk)

DOI: 10.4324/9781003571544

Typeset in Galliard
by Newgen Publishing UK

# CONTENTS

# FOREWORD

> Few tragedies can be more important than the reduction of life's possibilities, few injustices can be more profound than the refusal to provide the possibility to try or hope due to limitations imposed from outside and falsely identified as belonging to the individual.
>
> *(Gould, 1996)*

> Variations between individuals provide the platform on which the evolutionary process acts.
>
> *(Dunbar, 2003)*

> … As we convene at the World Economic Forum's Annual Meeting in Davos to discuss The New Global Context, the potential for elevated geopolitical and societal volatility shows that urgent action is required, particularly with respect to addressing local and global inequalities.
>
> *(Weber and Ermotti, 2015)*

> … fundamental, universal animal drives, such as dominance/prevalence, survival, kinship, and 'profit' (greed, advantage, whether of material or social nature), provide the basis for the evolutionary trap that promotes the unstable, conflictive, dominant-prone individual and group human behaviours.
>
> *(Colombo, 2022)*

## References


Colombo JA. 2022. *Dominance behavior: An evolutive and comparative perspective*. Springer Nature.

Dunbar RIM. 2003. The social brain: Mind, language, and society in evolutionary perspective. *Ann Rev Anthropol* 32: 163–181.

Gould SJ. 1996. *The mismeasure of man*. WW Norton.

Weber AA, Ermotti SP. 2015. *Prologue to The new global context: Could economic transformations threaten stability?* UBS White Paper for the World Economic Forum, Davos, Annual Meeting.

# INTRODUCTION

> In modern terms, the evolutionary trap has exchanged physical dominance for a cognitive gap, military dominance, and excessive wealth accumulation transformed into social power, media dominance, and political construction—a sort of privileged financial caste system. It is a web that pretends to keep molding and building our collective present and future towards the most profitable goals.
>
> *(Colombo, 2024)*

Living organisms emerged following a, so far, untestable origin of simple, primal molecular/cellular associations forms that colonized early Earthly substrates and surviving unimaginable, demanding ecological events. At some point, natural events provided grounds for the origin of continuously adapting and evolving life forms (definable as an organized set of biological structures able to adapt, survive, and multiply their living units) under various substrate and climate conditions. Thus, the emergence of micro-species was linked to such primal, changing environmental conditions.

The reformulation of new sets of macrobiological species following a series of adaptation failures due to geoclimatic events has been a repeated event that evolved through a sequence of geological eras. Before each of them, there were creatures of variable complexities adapted to different ecosystems and roaming on a planet undergoing continuous/periodic ground and climate changes. The survival cost was too high for uncounted terrestrial, water, and aerial species entering a palaeontological remnant phase. Yet this process allowed the emergence of fauna and ecosystems under different climate, fluid, and soil parameters. Natural history is built by species that lived and failed to survive at different rates. Within this unstable series of events, living forms of macro-species survived and later unable to survive in a continuous interaction involving their adaptive span capacity to variations in the ecological system. No species had a *firm living contract* in a changing world plagued with competitors thriving to adapt and survive. So, species emerged, flourished, and

DOI: 10.4324/9781003571544-1

disappeared to join the remains of unaccountable ancestors. Only a few species were able to adapt and prevail in a competitive world.

Yet were all natural history specimens, survival efforts, and experiences useless and drowned in a blind cesspool? And, closer to our interests, has the human species been built with the interactive inclusion of all that—yet almost unaccounted—precise history? Is our history as *Homo sapiens* another *brick in the wall* of natural history? Though the answer would be affirmative, humans have not yet assumed the implications of such conceptually appalling natural history projected onto the construction of our species.

Interactive, adaptive survival strategies of species depend on the species' physiological constructs and the limits imposed by external demands. Precisely, it is the emergence of a variety of species that, in the long term, somehow assures the continuation of living forms in a changing ecosystem. What is the impact of humans on this *survival equation*? On a global view of human actions, are we providing an element that potentiates an ecological and living species equilibrium and survival, or are humans acting blindfolded to its impact on Earthly universal survival conditions? From such a perspective of shared, interactive species survival, from an evolutive viewpoint, do human drives and actions belong to the *wrong species* (Colombo, 2010), since humans follow their goals irrespective of their environmental and ecological consequences? Means for survival built a natural history of strategies since living forms first emerged, as commented in the following paragraph from Winans and Bassler (2002):

> Several hundred millennia ago, prehistoric humans learned that there is often strength in numbers. The evolution of speech likely aided in the ability of protohumans to coordinate the behavior of the group. It has recently become clear that bacteria made this discovery at least a billion years earlier, and in lieu of speech, they evolved a rich lexicon of diffusible chemical signals. In the past decade, we have learned that bacteria use these signals to communicate both within and between species, and they also use a wide variety of molecules, signal detection apparatuses, and signal transduction mechanisms.

During its early times, *Homo sapiens* shared the risks and advantages of being one more species *Homo* among several in the balance of biodiversity. For approximately 15,000 years, the species *Homo sapiens* managed to out-compete other species and survive. While other *Homo* species, which were possible direct competitors for resources with *Homo sapiens*, began a regressive process, *Homo sapiens* began to spread around the planet and adapt to local demands. As time passed, *Homo sapiens* initiated a distorted relationship with the environment until it affected all ecosystems and accelerated the extinction of some species or increased the survival risk of others while promoting alternative life forms. This was until it became aware that its powerful enemy was reflected in the water mirror. The opponent was formidable, so it progressively developed ominous instruments of destruction and defence of ever-increasing power.

Humans today find disturbing perspectives complicating their vision of the shared future; hence, a manifest relativism arises in human society's supposed status as a 'dominant species'. Humans are in shared, conflictive control of a broad environmental niche but are limited by niches of the microworld in which other species are the dominant ones.

To place an evolutionary perspective on human behaviour as a species, a central notion should be mentioned from the beginning. That is, the evolution of *Homo sapiens* occurred within a radiation configuration—as opposed to a pyramid—of which palaeoanthropology is recovering some of its anatomical, functional, and molecular traces to build approaches to the biological identities of different variants of the genus Homo. The data obtained from recent genetic studies point out that *hominids* would have exchanged genetic material before and after the emergence of *sapiens*. Hence, from a phylogenetic point of view, *Homo sapiens* would not have arisen from a line genetically isolated from previous and cohabitant close congeners.

One diverging point from the primate evolutive line—perhaps related to adaptive evolution or positive selection—was represented by a developmental characteristic of the human brain, which continues to grow for a long time after birth (Coqueugniot et al., 2004; DeSilva and Lesnik, 2008) and its basic organization—fundamentally involving cortical circuits—is not completed until reaching adolescence. This implies a high degree of initial neurobiological immaturity and helplessness that requires significant containment mechanisms during the developmental period, very much longer than the other mammalian species. The neuroplasticity involved allowed the human species to overcome such a competitive handicap, suggesting the emergence of a positive evolutionary interaction of long-term postnatal biological and social factors.

The highly cherished collective ego of humankind has somehow succeeded in hiding culturally the ancestral, evolutive human origin as part of natural processes and events, perhaps by implementing two main strategies. They would be, first, by attempting to downplay natural biological and evolutive evidence, and second, by implementing cultural or cognitive constructions moulded to caress the human collective ego and to differentiate humans from the surrounding natural context, appealing to mystical constructions. The conceptual conclusion under such subjectively conditioned thinking is that we are some extraneous implants in natural history with the capacity to (and the right to?) exert dominance upon natural species and modify the ecosystem to meet our changing human needs. The cultural origin and evolution of this distorted, self-oriented view of human biological construction may have its origin following millennia of interactive struggle between the limitations of the human biological condition and the display of natural events. The latter tended to result in the construction of a submissive self-conscience expressed under multiple cults and religious constructions throughout human history. These were bound to submissive behaviour towards uncontrollable physical events, often announced, and their interpretation manipulated, by psychologically dominating human actors and esoteric constructions that promised relief from Earthly bound disgraces. Thus, collective submission was transferred from natural events to human actors and their virtual constructions.

In such regard, according to Dein and Littlewood (2011), religious cognition and schizophrenia recruit related mental modules linked to the oversizing of acting entities and the *theory of mind*. This concept was advanced by Polimeni and Reiss (2002), who stressed similarities between shamanism, religion, and psychosis (also discussed by Thalbourne and Delin, 1994). In such a regard, the psychotic condition could have been manipulated by shamanic leaders who incorporated novel rituals based on their altered state of consciousness and hallucinations. Other manifestations must have been socially shared events and affected naïve and overwhelmed minds concerning natural life and death events from ancestral times. The construction of esoteric interpretations and mystic powers would have needed the co-optation of permissive brain neural circuits. These behaviours would have included emotional imbalances, delusions during febrile states, shared experiences of dreams, or involved the consumption of herbs with psychoactive effects. These ancient behavioural profiles, expressed under different formats, must have resulted in an active, effective combination of the mental conscious and subconscious organization during an 'early adolescence' of the human species that has persisted until nowadays.

A brief survey of human anatomical construction will rapidly find remnants of what belongs to the evolutive natural history of the *Homo sapiens* species. Let us mention a few of them—for example, the remains of anatomical segments such as those found in the postnatal coccyx segment of the vertebral column (a foetus with a caudal appendix or tail during embryological stages 14–22, followed by partial reabsorption); auricular muscles attached to our ears that allow auricular orientation; the remnant appendix; and branchial pharyngeal folding. In addition, the integrated function of various organs and regulatory mechanisms depend on human physical and mental health, reproductive capacity, and survival strategies moulded by a brain—a plastic, multipurpose, and interactive organ that integrates physical and emotional events; a brain where thoughts and introspective and exteroceptive perception and behaviour are integrated as part of the construction of human identity. In this regard, the existence of sensorial and perceptual processes is common to living species, and their physical capabilities are characteristic of each, evolved through millennia of interactive existence with the surrounding environment and its specific physical niche.

Through the millennia of evolution, the neural component—an integrated conglomerate of receptors, information processors with a final common internal and external behavioural path, and intrinsic neuroplasticity—has acquired increasing complexity involving communication codes and semantic richness. This increasing neural complexity allowed humans to constantly reformulate self-representations of plastic, interactive models of the surrounding reality, and trigger behavioural adaptations.

Each species evolved its developmental characteristics subjected to genetic and environmental cues. In *Homo sapiens*, the extended postnatal period of the cerebral organization (neoteny) made possible the extended period of social shaping of the mind from an early age—socialization/domestication—and adjustments through the learning of our complex social behaviours. Behavioural, gregarious profiles allowed for different configurations of social groups and conditions for early brain/mind development. In this regard, Charles Darwin (1943) stated:

> The main conclusion we reach in this work, that is, that man descends from some inferiorly organized form ... There would hardly be the slightest doubt in recognizing that we descend from barbarians.
>
> *(p. 791)*

> ... man ... bears in his corporeal make the indelible seal of his inferior origin.
>
> *(p. 721)*

> Therefore, we can infer ... that if men are in decided superiority over women in many aspects, the average mental faculties of men will be above those of women.
>
> *(p. 721)*

These comments adhere to inappropriate adjectives and judgements regarding species differences, considering that each species represents the best transient adaptation to their ecological environment. However, they do not obscure Darwin's enlightening critical findings and general concepts on species evolution. Yet, in natural sciences, there should be no room to include comparative value judgements on species-specific evolutive, individual environmental adaptations, physiological profiles, and intrinsic (plastic) trends within the universe of the natural kingdom, as expressed, probably linked to cultural, historical domains.

Emphasis should be placed on the evolution of language, which allowed the development of species-specific complex structures of thought and collective behaviours of increasing sophistication potentiated by abstract thinking. This summarized view of increasing evolutive brain/mental and behavioural complexity would only be complete by including the concept and interactions of the manifest (conscious) and subconscious dimensions, the first involving the concept of *working memory*. In this regard, a significant component of cognitive processing appears to be at the subconscious level; what seems at the conscious level, expressed temporarily at a specific time, are the events we can manipulate as a working memory in our executive behavioural domain that would correspond to the explicit memory of Dietrich (2015). One main aim of this book is to provide additional views to understand better the construction of sociality and the potential risks that would depend on ancestral territorial and prevalent drives imprinted in the human brain and behavioural trends that impinge upon current social behavioural drives and interactions. This is because fundamental human nature did not emerge from a *tabula rasa* condition but from previous evolutive events that included a spectrum of genetic pool exchanges, climate, geophysical conditions, and early conditions of prey and predator. From such a multifactorial equation emerged—now extinct—several *Homo* species until the birth of *Homo sapiens*, an expansive, predominant macro-species. This resulted in the emergence of the *human factor* in geological and natural events.

Though predation and cooperation represent constant behavioural profiles in natural history, a dynamic equilibrium lasted until the emergence of the master predatory species, for *Homo sapiens* has placed at risk the equilibrium of natural ecological variables. A dominant species that evolved cultural and technological strategies resulting in the over-exploitation of natural resources, development of arms of massive destruction, strategies to foster massive consumerism, and political means to manipulate public opinion. This combination represents a powerful means to exert dominance and enter a competitive race towards it. As a result—besides the risks of warmongers attempting *dominance*—poverty, deprivation, marginalization, and slavery work represent a crystallization of prejudices that emerge from social dystopias supported by a financial/mercantile view of human relations.

The emergence of the human species with a sophisticated language allowed the development of complex virtual constructs based on symbolism and the instruments of culture, which has enhanced cognitive capacity and emotional interaction. However, all of them are supported by processes anchored in neural networks distributed in an interactive mode within cortical and subcortical levels, with essential neurobehavioural components preserved during the species' evolutionary development. As mentioned before, they refer mainly to the hedonic attractiveness of power or prevalence, securing territory, dominance, and reproductive and food rights, which have become part of the modern cultural process in competition, consumption, accumulation, violence/belligerence, personal contribution, narcissism, and privileges—all generators of inequities.

Thus, our *sapiens* species replayed with increased efficiency and new behavioural profile categories universally branded as prey, predator, and scavenger. Modern humans have found new strategies and cultural forms to disguise such social categories. These asymmetries conspire against *globalization*—understood as the functional integration of humanity and not only of the 'financial markets' or comparatively privileged segments—and instead they also seem to promote social segmentation regarding access to rights and education and cultural prevalence.

In parallel to astounding developments, few minorities promote positive behavioural characteristics, represented by creativeness, solidarity, and social care. This underlying sociocultural current represents the core treasure of the human species that could seed a different future. In fact, according to Martindale (1999) and Carson (2014), creativity has been associated with cognitive and behavioural disinhibitory processes. It ought to be considered, though, that inhibitory mental processes are not only fundamental to rational behaviours but also involved in socially adaptive practices. Hence, promoting creativity must overcome the limits imposed by diverging socialization processes. An undue imbalance of inhibitory mental processes and socioeconomic domains keeps tensing human populations' profiling and divergent futures.

Due to the dynamics of the regime of consumption and exploitation of natural resources, the development of powerful technologies, and the factional struggle for power today and in the future, a managerial responsibility has been raised that exceeds the local well-being of fellow citizens. These involve living conditions and survival and viability on the planet. In this sense of current and potential damage, 'globalization' is operative—it is imposed—but, unfortunately, it is in a vicious context: universalizing the progressive degradation of the planet's natural history and the environmental living conditions on it and avoiding through social/financial dominance the expression of diverse cultural drives.

The foregoing text supports the thesis that without education and the promotion of universal values involving individual opportunities to evolve and protect the environment,

there will necessarily be a further decoupling of the integration of communities. This fracture will deepen, given the speed of social behaviours and technological advances that are underway. The reality is that there are significant pockets of communities that are undernourished, impoverished, and without access to primary healthcare or adequate education for the continuous changes in the modern world, and—essentially—enslaved at the service of corporations or wealthy and dominant power groups.

## References

Carson SH. 2014. Cognitive disinhibition, creativity, and psychopathology. In DK Simonton (editor). *The Wiley handbook of genius* (Chapter 11). Wiley. doi.org/10.1002/9781118367377.ch11.

Colombo JA. 2010. *Somos la especie equivocada?* (*Are humans the mistaken species?*). Eudeba. https://tinyurl.com/lsdozicg.

Colombo JA. 2024. *Evolution and the human-animal drive to conflict. A psychobiological perspective.* Routledge.

Coqueugniot H, Hublin J-J, Veillon F, Houet F, Jacob T. 2004. Early brain growth in Homo erectus and implications for cognitive ability. *Nature* 431:299–302.

Darwin C. 1943. *El origen del hombre y la selección en relación al sexo.* (*Human origin and sex related selection*). Albatros.

Dein S, Littlewood R. 2011. Religion and psychosis: A common evolutionary trajectory? *Transcult Psychiatry* 48: 318–335.

DeSilva JM, Lesnik JJ. 2008. Brain size at birth throughout human evolution: A new method for estimating neonatal brain size in hominins. *J Human Evol.* 55: 1064–1074.

Dietrich A. 2015. *How creativity happens in the brain.* Palgrave-Macmillan. doi: 10.1057/9781137501806.

Martindale C. 1999. Biological bases of creativity. In RJ Sternberg (editor), *Handbook of creativity* (Chapter 7). Cambridge University Press.

Polimeni J, Reiss JP. 2002. How shamanism and group selection may reveal the origins of schizophrenia. *Medical Hypotheses* 58: 244–248.

Thalbourne MA, Delin PS. 1994. A common thread underlying belief in the paranormal, creative personality, mystical experience and psychopathology. *J Parapsychol* 58: 3–38.

Winans SC, Bassler BL. 2002. Mob psychology. *J Bacteriol* 184: 873–883. doi: 10.1128/JB.184.4.873-883.2002.

# 1

# BIOLOGICAL AND CULTURAL DEVELOPMENT OF *HOMO SAPIENS* AND CULTURAL CONDITIONERS

To place the human species on an evolutionary perspective, a central notion should be mentioned from the beginning: the evolution of *Homo sapiens* would have taken place within what configures a radiation—as opposed to a pyramidal shape—of which palaeoanthropology is progressively recovering some of its anatomical, functional, and molecular traces to build estimates of the biological identities of different variants of the genus *Homo*. The data obtained from recent genetic studies point out that genetic material was exchanged among hominids before and after the emergence of anatomically modern man. In such a way, from the phylogenetic point of view, humans would not have arisen from an evolutive line genetically isolated from members of the evolutive *family* of *Homo* species nor from one that was alien to mammalian emergence. Such would be the case of the possible contribution of Neanderthals and Denisovans—considered *archaic Homo sapiens*—to the genetic pool of the human genome (Harmon, 2012), as well as previous *Homo* species intermixing.

A differential comparative characteristic of *Homo sapiens* among *Homo* representatives—except for some species of 'archaic' *sapiens*, like the Neanderthals—is the development of brain volume and its neurobiological characteristics, mainly due to the significant expansion of its cerebral cortex, as expressed by the increased cranial capacity. Besides cranial capacity, the biologically degradable evolution of connectivity changes and structural and molecular domains does not allow for early fossil research, thus only allowing comparisons limited to anthropometrics or craniometric estimates with our fossil ancestors and some functional inferences. In this regard, using samples from surviving species, a possible relationship with the socialization process and its growing complexity that took place within the primates and particularly in the genus *Homo* has been attributed to the relative increase in cortical grey matter—and its structural and dynamic characteristics of neuronal density and connectivity (see, for example, in Sallet et al., 2011; Dunbar, 2003; Flinn et al., 2005).

The complete reconstruction of habits and capabilities of extinct *Homo* species is limited by the absence of records of its possible behaviours, of which only traces have been

DOI: 10.4324/9781003571544-2

left based on their habitats, the production of instruments, shared remains with other specimens, traces in the fossil soil composition suggesting possible eating habits, gender body dimorphisms, and comparisons with surviving species of primates. It is conceivable that certain evolutionary components (e.g., adaptive evolution or positive selection) must have influenced such brain development, inasmuch as the human brain undergoes a significant, critical, postnatal growth (Coqueugniot et al., 2004; De Silva and Lesnik, 2008), and its essential organization—fundamentally regarding soft- and hardware cortical circuits—is not completed until past the time of adolescence.

On behavioural and self-sufficiency grounds, this implies a marked degree of initial neurobiological immaturity and helplessness that require containment mechanisms for the *sapiens* species during postnatal development, significantly expanded beyond that of the other primate and general mammalian species. To allow for this extended postnatal period of brain/mental processes has indicated that the human species had to overcome such a competitive handicap, which is only possible due to the emergence of a complex evolutionary interaction of biological, behavioural, and social factors.

A cerebral energy consumption of close to 80 per cent of the total body at rest during the early postnatal period (Leonard and Robertson, 1992, 1994; Leonard et al., 2003) indicates the degree of such demand and the significant survival fragility of that period, most notable during early evolutionary stages. This developmental characteristic implied its dependence on an early, adequate, and complete diet and physical protection. These demands imposed since the germinal times of the species would have required an influential community organization in the face of natural risks imposed on child survival.

Establishing the 'private domain' concept is deeply anchored in the evolution of animal—and plant—behaviour, later conceptualized by humans as 'private property'. The extension of this survival territory will depend on the energy requirements of the leading individual of the species to which it belongs and of its social group, its tactics, its attack and defence capabilities, and the characteristics of herd behaviour. Through evolutive horizons, physical (material) possessions (objects, territory) and normative (hierarchical) cultural profiles conditioned the expression of this primitive, collective behaviour that is included in modern human behaviour. Following eons of evolutionary trial and error, *Homo sapiens* incorporated fundamental additional strategic, material, and technological elements, added to the standard set of behavioural parameters of animal species. It included a sophisticated domestication (socialization) process aimed at coupling the individual to their community—which is expected for gregarious species behaviour.

Regarding the comparative use of material elements, although the instrumental use of natural objects has been described in different species (birds, non-human primates), the progressive complexity of craftsmanship achieved initially by the carving of tools (arrowheads/spears) established a progressive dominance probability. Further, the pictorial expression of surrounding elements by those who preceded modern humans in the *Homo* lineage indicated the emergence of progressive forms of historical record. These seminal behaviours mark the divergence among animal species in the complexity of the processes involved in intra- and extrapersonal behaviour. The emergence of symbolic language potentiated logic and rational processes that anticipated the subsequent positive slope of scientific-technological development.

From poachers, gatherers, and scavengers, humans progressively began to control the energy yield of the land and its biodiversity, and with it came the settlement, the progressive demographic explosion, and increasingly complex social organizations that channelled further competitive encounters based on individual or communal possessions. Without being fully aware of it initially, *Homo sapiens* began to compromise the *ecological footprint* (a measure of the demand imposed by land and sea production on the Earth's ecosystems), that is, the limits of the mineral and biological resources of the planet, to satisfy food, productive, and energy consumption demands. The following statement by Dinerstein et al. (2024) represents a call for conservation imperatives:

> In late December 2022, at the United Nations Convention on Biological Diversity's 15th Conference of Parties (COP15), more than 190 parties adopted the 30{x}30 target—to protect at least 30% of the world's lands, oceans, and inland waters by 2030. Conservation biologists, Indigenous Peoples, science-based NGOs, corporate leaders, and others have endorsed the 30{x}30 target and also called for protecting half of the terrestrial realm for humanity to have the best chance to reverse biodiversity loss, stabilize Earth's climate, prevent ecosystem collapse, and avoid future pandemics.

Within several years, the exploitation of natural resources was promoted, and the bases of mystical and religious thought, the emergence of alternative idolatries, written language, artistic expression, and later scientific thought were shaped and differentially embraced the behaviour and beliefs of extended human populations. In this regard, in recent years, technology, 'the market', and consumer behaviour have promoted different ways of dealing with a complex reality tinged with identity profiles, emotions, interests, assumptions, and prejudices.

On emotional grounds, it involved the expansion of mystical and religious drives that emerged anchored to feelings of fear and submissive behaviour towards the unknown. These were potentiated by the lack of objective understanding of the natural phenomena and of a natural process distorted by beliefs about the termination—death—of the individual being. The unknown or the 'hidden' has always motivated fantastic interpretations, rationalized in different ways, and shaped various cultural profiles. In this regard, the comments made by Polimeni (2012) are interesting:

> The polytheistic religions of traditional societies undoubtedly share certain similarities to contemporary monotheism: both deal with supernatural entities and include rituals concerned with food procurement and life transitions such as birth, marriage and death ... hominid warfare may have created the greatest selection pressures guiding the development of shamanism and religion.

*Homo sapiens* species development took place in a natural evolutionary framework. However, cultural profiles and mystic processes generated proposals that attempted to isolate it from the ***natural kingdom*** and build its emergence as an autonomous event, detached from natural history. Thus, since the initial stages of the species, it progressively built a condition receptive to the construction of parallel, virtual worlds detached from the natural, material world and inhabited by virtual agents with fantastic powers, progressively attempting the rationalization of these beliefs. Constructing an imaginary parallel world aided in building emotional security amid physical insecurities and the notion of the finiteness of living.

Magical thoughts progressively culturally invaded the 'emotional mind' of the new dominant species, and from there, it was projected to involve various dimensions of individual and social behavioural profiles. Some expressions of altered mental states established emotional bridges with all-powerful occult powers in situations of famine, war, or death threat. Rigid social structures promoted the emotional mobilization of faithful servants/followers in pursuit of beliefs, sometimes transformed into fundamentalisms operated from the highest hierarchies of the social organization. These emotionally bound behaviours generated leaders who, in the collective imagination—or autocratic systems—acted as hypothetical bridges with 'the world of the unknown'—a 'superior world' in contrast to the current Earthly one. Since then, and possibly involving a projection of what constitutes an intrinsic, basic behaviour for the survival of organisms, the *Homo* world was progressively transformed into an abusive, unequal entity with profound asymmetries—some of which led to processes of partial extinction—an adjourned Darwinian replica of the survival dynamics.

With the emergence of cognitive expansion and emotional expression, the original prevailing mystical/religious constructions established that the imagined certainties about the inscrutable territory of death—and the gods involved—carved deep into self-consciousness and developed a mystical/religious culture where access to *truth* and paths towards mystical constructions were only accessible, while alive, to a select few. For the 'commons', their personal, uncritical dedication to religious practices offered safe conduct towards emotional self-contempt and the proclaimed afterlife. This adherence—belief, faith—to a supposed posthumous award promoted both submission and sublimated expressions of creativeness—especially in the arts and letters—and the construction of a universal ethical framework.

Later, movements of thought—religious or not—began to claim 'human rights' for the dispossessed or socially relegated in the distribution of rights or social welfare and the need to redistribute Earthly wealth. Some were part of religious doctrines that progressively dissociated man's attainments in everyday living and the promised afterlife. That is, the Earthly psychological dimension and *Homo* ancestral fears promoted the build-up of actual empires of faith that tended to sublimate those fears and place them under an umbrella of mystic constructions. Thus, behaviours were moulded based on 'directed altruism', which varied according to religious beliefs, rules, and social norms of the time, place, and cultural history.

On Earthly grounds, the material inheritance of their leaders often increased in parallel with their power to manipulate the fears and insecurities of their followers. Today, there are religious corporations with valuable assets, while numerous legions of their human flock drag through famine, marginalization, and poverty, embraced by mystic notions.

Supported by their creativity and the desire to overcome the limits of memory, groups of leaders of *Homo sapiens* began to build the foundations—sometimes monumental—of their belief. This is often based on a mental trap camouflaged behind rites, promises of eternity and redemption, and a magical relationship with natural or unknown phenomena.

Due to the advent of farming settlements and land exploitation, progressive land yields evolved into an increasing economic, social, and political power source. They progressively triggered the concept of private property as a financial asset and the emergence of new social categories. In this realm emerged new categories, such as the conceptual development of 'poverty' as an economic category (Colombo E, 2007). Property inequalities and richness probably emerged following the end of the *Neolithic* period associated with land property in Eurasia. They increased farming and domestication of livestock, as has been dealt with by Kohler et al. (2017). The changing concept of dominance among humans from the original animal concept has represented an evolutionary inheritance that significantly primed human history, adding a central element in constructing hierarchical societies and social inequalities, and political or financial oppression.

This complex duality between the physical and the emotional drives impregnated the vision of the limits of man's freedom. It occurred even considering the relative nature of such freedom and its dependence on its biological nature and evolutive circumstances. That is, based on the interaction of our genetic menu and our culture. From this plot, it becomes difficult to extract significant quotas of autonomy of free thought other than those thoughts limited or contained within those same constructions. This notion cannot be reviewed without considering *ab initio* human construction of the concept of 'freedom' or, perhaps more accurately, of our 'degrees of freedom' (Colombo JA, 2013) and human self-awareness of the internalization of inhibitory codes and rules in the socialization process.

This author finds this process too close to domestication and stifling creativity. However, sublimated behavioural and creative–artistic–expressions of beauty and social behaviours of true personal devotion emerged from that ambiguous human heart of omnipotent thought relative to nature and submissive to the notion of death or guilt rituals.

Under the protection of faith and its emotionally bound constructions—religious or not—the human mind has been capable of generating genuinely unusual transformations, whether under—or despite—pain and misery, or based on the comfort of patronage, or from the unequal struggle to sustain an ideology irrespective of oppressive conditions. Giordano Bruno's critical thought transformed into a burning pyre, Leonardo's soft brushstrokes, Michelangelo's delicate communication using blocks of marble from Carrara, Galileo's final forced attempted muted conviction during his trial, Spartacus' dedication to freedom and his slave followers, Madame Curie's self-immolation for the sake of knowledge, or Ghandi's dedication to his cultural revolution—these are all examples of transcendent behaviours that provided sublime human values (creativity, beauty, solidarity, equity), and have been recognized as such by different generations. Their historical commentators will be able to relativize and contextualize their behaviour. Still, some actors will remain trapped in the fabric of collective memory, even if times of history change. A movement for justice, an allusion to beauty, or a collective rebellion will suffice to revive them—their symbolic image or central contribution.

Religious faith, its doctrines, and its ethical framework have added value to social behaviour. But that key often also accesses emotions that transmute into fundamentalist, dominant behaviours that reduce freedom of thought to the point of triggering oppression, wars, sentences to cruel physical or mental torture, and what has been called the *manufacture of consent.*

Dominance behaviour (Colombo JA, 2022, for further analysis) is expressed in different ways by all species, and it contributes to setting the basis of social organization and the ecological balance within animal ecological niches. Regarding the differences or asymmetries in living conditions and individual/community development possibilities, it is enough to look at the data on inequalities on the planet. While *Homo sapiens* initially played within the rules of the game of *natural history*, the Darwinian relationships of humans were progressively expressed in a framework of balance between prey and predator/gatherer within the limits imposed by the ecosystem. Developing human skills led to slow population increases with periodic crises of confrontation and migration episodes. Both became more marked when they began to control the yield of the land and progressively detach from the risks of the prey-predator relationship and limits imposed by the ecosystem. The Darwinian relationships exceeded those directed at the ecosystem and were increasingly concentrated in actions within the same species, invading the economic, cultural, and war-prone dimensions. As dealt with by Elhacham et al. (2020):

> Humanity has become a dominant force in shaping the face of Earth. An emerging question is how the overall material output of human activities compares to the overall natural biomass … in the year 2020 (± 6), the anthropogenic mass, which has recently doubled roughly every 20 years, will surpass all global living biomass. On average, for each person on the globe, anthropogenic mass equal to more than his or her body weight is produced every week.
>
> At the beginning of the twentieth century, anthropogenic mass was equal to only 3% of global biomass, ... About 120 years later, in 2020, anthropogenic mass is exceeding overall biomass in the world.

By 2023, the population exceeded eight billion human inhabitants on Earth. The waste from its activity accumulates in the Pacific Ocean, on lands of forgotten fertility, in aquifers, in atmospheric puffs, and in the relative solitude of nearby extraterrestrial space (scrap from tests—and trips—following the installation of the space station and communication/research satellites). Little has been left untouched by the ambition of developers, corporations, and political powers, though a large world population lives on trash and leftovers. It is as if there are 'strains' of humans with excessive ambitions, compromising the world's survival or quality of life by directing excessive resources to other means than people's well-being and pursuing primal search behaviour.

Coping with internal and external demands constitutes an essential characteristic of species' behaviours. It aims to reduce the mismatch between environmental demands and physiological limits for adaptation and acquiring a dynamic equilibrium condition, i.e., a potentially stationary condition of instability. This involves behavioural responses such as aiming for shelter or migration. These complex, interactive sets of variables acquired further complexity among humans regarding social rights.

In ontogenetic evolutive terms, the transitions to extrauterine life represents a formidable leap for human development that requires appropriate physiological responses to the imposed demand. This poses a significant demand to cope with the 'leap' implied in changing from a physiologically cushioned, conditioned environment to a sudden entry into an environment that demands the continuous adjustment of variables that must be activated and linked to vital parameters of the internal environment. Under normal conditions, for an extended period, the interaction with the mother provided the bridge between those needs and their achievement, constituting the first socialization lesson. Thus, the relative abrupt change from a conditioned environment to the extrauterine environment represents a complex set of variables and interactive physiological and behavioural adjustments that probably have a long-term imprinting effect.

> An infant's first postnatal experience is satisfying the need for food at the breast and the proprioceptive sensation of warmth from the mother. With or without crying turned into an instrument, the achievement is the hedonism of the satisfied need. And this configures the first success of social interaction. That is, the first learning is based on pleasure.
>
> *(Colombo JA, 2019)*

This first experience is also a conditioner of self and third-party behaviours. The initial role of the mother as an unconditional provider constitutes the first and most resounding experience of social altruism for the newborn citizen and the exercise of a self-profit, indulgent experience. The consequence is that by 18–20 months of age, the child has consolidated a behavioural marker experience based on hedonic and selfish profiles. As Warneken and Tomasello (2008) commented, the child does not behave selflessly by natural tendency, for it grew up building on such a vital experience phase and will express it with the means at its disposal until it enters the interactive social context.

From comparative anthropological and evolutive considerations, we can see that humans have a very extended period of postnatal dependence during which the basic scaffolding of their brains and mental development is completed, and that this rearing period imposed demanding conditions for the emergence of *Homo sapiens.* Given palaeoanthropological data available, even in related species of the genus *Homo*, or other primates, the period of postnatal brain growth and behavioural immaturity is shorter than in *sapiens.* The result in these species is a more limited childhood and brain volume, which is therefore not equivalent to that of the *sapiens* child (Colombo JA, 2009, 2019).

> After millennia and centuries, the 'Darwinian jungle' changed actors, but its essence persists. Given the fragility and plasticity of the infant brain, and the defencelessness of the infant, the risks for its offspring remain in force. But now, it is subject to variables generated by the human being himself: culture. In him, upbringing is a behavioural characteristic modulated by cultural dynamics. This—as well as the full exercise of parenting– is significantly affected by strategies and decisions linked to public policies that often result in marginalization, exposure to unhealthy conditions, lack of biomedical containment, lack of food and education, etc. These conditions carry the risk of brain and mental damage of social origin.
>
> *(Colombo JA, 2019)*

Mutations and epigenetic changes—heritable changes in DNA expression due to modifications of nuclear chromatin– evolved since the very origin of life in its most elementary forms. This would have allowed more complex organizations to reach adaptable or metastable equilibrium conditions. This internal equilibrium and adaptation to environmental conditions was conceptualized by Cannon (1926) as homeostasis. It implied the continuous adjustment of variables linked to vital parameters of the internal environment (electrolytes, hormones, temperature, blood glucose, etc.) within the limits of survival, specific for each species. This also involved behavioural adaptations, using which individuals of the animal kingdom survive by adjusting the variables of their biological and behavioural parameters to minimize extreme variations.

## References


Cannon WB. 1926. *Physiological regulation of normal states: Some tentative postulates concerning biological homeostatics.* Paris: Éditions Medicales.

Colombo E. 2007. La pobreza como categoría económica (Poverty as an economical category). In JA Colombo (editor), *Pobreza y desarrollo infantil* (*Poverty and child development*). Paidos Editorial, pp. 33–51.

Colombo JA. 2009. Desarrollo infantil: Consideraciones evolutivas y neurobiológicas para políticas públicas (Child development: Evolutionary and neurobiological considerations for public policies). In *Derechos vulnerados en la infancia: Abandono, maltrato y pobreza* (*Vulnerated rights in childhood: Abandonment, abuse and poverty*). Editorial de la Universidad Católica Argentina, pp. 103–109.

Colombo JA. 2013. *Bajo libertad condicionada* (*Under conditioned freedom*). Imago Mundi.

Colombo JA. 2019. *Our animal condition and social construction.* Nova Science Publishers. ISBN: 978-1-53615-357-6. eBook ISBN: 978-1-53615-3583.

Colombo JA. 2022. *Dominance behavior: An evolutive and comparative perspective.* Springer Nature.

Coqueugniot H, Hublin J-J, Veillon F, Houet F, Jacob T. 2004. Early brain growth in Homo erectus and implications for cognitive ability. *Nature* 431: 299–302.

De Silva JM, Lesnik JJ. 2008. Brain size at birth throughout human evolution: A new method for estimating neonatal brain size in hominins. *J Human Evol* 55: 1064–1074.

Dinerstein E, Joshi AR, Hahn NR, Lee ATL, et al. 2024. Conservation imperatives: Securing the last unprotected terrestrial sites harboring irreplaceable biodiversity. *Front Sci* 2: 1349350. doi: 10.3389/fsci.2024.1349350.

Dunbar RIM. 2003. The social brain: Mind, language, and society in evolutionary perspective. *Ann Rev Anthropol* 32: 163–181.

Elhacham E, Ben-Uri L, Grozovski J, Bar-On YM, Milo R. 2020. Global human-made mass exceeds all living biomass. *Nature* 588: 442–444.

Flinn MV, Geary DC, Ward CV. 2005. Ecological dominance, social competition, and coalitionary arms races: Why humans evolved extraordinary intelligence. *Evol Hum Behav* 26: 10–46.

Harmon K. 2012. New DNA analysis shows ancient humans interbred with Denisovans. *Nature.* doi 10.1038/nature.2012.11331.

Kohler TA, Smith ME, Bogaard A, Feinman GM, et al. 2017. Greater post-Neolithic wealth disparities in Eurasia than in North America and Mesoamerica. *Nature* 551: 619–622.

Leonard WR, Robertson ML. 1992. Nutritional requirements and human evolution: A bioenergetics model. *Am J Hum Biol* 4: 179–195. doi: 10.1002/ajhb.1310040204.

Leonard WR, Robertson ML. 1994. Evolutionary perspectives on human nutrition: The influence of brain and body size on diet and metabolism. *Am J Hum Biol* 6: 77–88. https://onlinelibrary.wiley.com/doi/abs/10.1002/ajhb.1310060111 1/2.

Leonard WR, Robertson ML, Snodgrass JJ, Kuzawa W. 2003. Metabolic correlates of hominid brain evolution. *Comp Biochem Physiol A Mol Integr Physiol* 136: 5–15. doi: 10.1016/s1095-6433(03)00132-6.

Polimeni J. 2012. *Shamans among us: Schizophrenia, shamanism and the evolutionary origin of religions.* Evo Books.

Sallet J, Mars RB, Noonan MP, Andersson JL, et al. 2011. Social network size affects neural circuits in macaques. *Science* 334: 697–700.

Warneken F, Tomasello M. 2008. Extrinsic rewards undermine altruistic tendencies in 20-month-olds. *Dev Psychol* 44: 1785–1788.

# 2

# EMERGENCE AND DEVELOPMENT OF *HOMO SAPIENS*

> A human and a grain of rice may not, at first glance, look like cousins. And yet, we share a quarter of our genes with that fine plant. The genes we share with rice—or rhinos or reef coral—are among the most striking signs of our common heritage. The percentage of genes we share with various other species is a clear indicator of our common origin.
>
> (Zimmer, 2014)

> This suggests that hominin evolution is neither a simple punctuated process, nor a constant gradual one, but a complex interaction between variable rates of change, environmental dynamics and the competitive interactions of the hominins and their sympatric[1] fellow-travellers in evolution.
>
> (Foley et al., 2016; note added by book author (JAC))

The emergence of *Homo sapiens* occurred as one additional species of *Homo* radiation that recognizes numerous trials adapted to different ecological niches. It was preceded by millions of years of evolution, a natural history governed by genetic composition, species competition, climatic changes, and progressive and catastrophic geological and climatic events. In this domain, the *Zoonomia* project (Vignieri, 2023) explores the joint production of genomes from hundreds of species that would aid in understanding mammalian evolution.

Species evolution undergoes a spectrum of alternatives in which humans represent a conditioned outcome built with materials from natural history. This was preceded by countless survival attempts of today's fossil species, where the dominant species of yesterday are archaeological fossils or conceptual constructions applied to attempt the reconstruction of animal history.

DOI: 10.4324/9781003571544-3

In this interactive process with conditioned probabilities and still unsolved admixtures, genetic exchanges among *Homo* lineages represent a formidable, hazy backbone of *Homo* evolution that evolved into current *Homo sapiens* dominance.

> Further studies of the phenotypic consequences of hybridization across taxa have shown that admixture of different lineages may have provided genetic variation that was then favoured by natural selection to drive evolutionary innovation in our hominin ancestors.
>
> *(Ackerman et al., 2019)*

Thus, the development of human species involved a series of genetic exchanges among successive *Homo* species, of which Neanderthal and Denisovan would have been the latest ones. As posed by Harris et al. (2023), genetic flow between Neanderthal and modern human genomes would have been bidirectional. Though human history has been attributed with an existence of about 200,000 years of existence, it eventually became the current 'dominant macro species' (Colombo, 2022).

In such a lengthy and aleatoric interactive process, the human species evolved a brain composed of an organization that recognizes different stages of evolution since the original *cephalic ganglion* of primal species. The newly evolving brain structures were progressively embedded in new connectivity, allowing new capabilities of brain processing, a behavioural domain that coexists with several basic regulatory components. These involve thermal balance, normoglycemia (appetite for intake), sexuality—reproductively oriented or not—, defence mechanisms (passive or active), and social and ground territoriality. The interaction between the cultural and the biological *domains* is part of the human ability to achieve individual and social balance, human behaviours, and human internal tensions—dissatisfactions and contradictions—as well as the continuous update of cultural interactions throughout history. The predominance—or mismatch—of one or the other is manifested in the various behaviours of the human species, whether creative, adapted, subdued, dysfunctional, or pathological.

According to Devièse et al. (2017) and Hublin (2017), data would show that all the Neanderthal remains are from an earlier period (>40,000 cal BP [the calendar year before the present]) than originally estimated.

> The latest data, both radiometric and genetic, suggest Neanderthals and modern humans coexisted or overlapped for up to several thousand years in Europe until Neanderthal disappearance at around 40,000 cal BP.
>
> *(Devièse et al., 2017)*

As *Homo sapiens* spread worldwide, they would have encountered Neanderthal settlements during an extended period (perhaps as much as 15,000 years), probably enough to provide opportunities for interactions and limited genetic exchange to take place until Neanderthal's disappearance as a species (Callaway, 2021).

Initially, *Homo sapiens* shared the competing risks and advantages of being another *Homo* species among several others in the balance of biodiversity. Over time, it managed to progressively distort its relationship with the environment to affect all ecosystems, accelerating the extinction of some species and increasing the extinction risk for others. Competition for survival among a variety of *Homo* competitors continued for some time until, at some point, it became evident that the human species' powerful enemy was itself. The opponent was formidable, so the species successively developed increasingly powerful and ominous instruments of destruction and defence. Today, humans find themselves with disturbing questions that complicate their vision of the path to travel: how to safely survive and interact with the planet's natural life and ecosystems and deal with cultural differences and dominance competition?

Scientific estimates (Mora et al., 2011) suggest that there should be close to 7 million terrestrial species and more than 2 million maritime species globally registered. Yet, many more species would remain to be identified. It is not only their number but their survival skills under different ecosystems that should make us think about humans' supposed exclusivity of resources and abilities. Many of these species have preceded humans and probably will succeed them, too, especially those biologically simple forms capable of rapid adaptive mutation (e.g., extremophiles).

Humans should become aware—should assume it culturally—of the existence of biologically complex beings that arose from what would have been two primary domains of life forms—*Archaea* and *Bacteria*. In this context, leaving aside precisely those initial forms of life, of all the taxonomic groups of the natural kingdom, insects—which would have originated some 390–350 million years ago—are the most abundant today, constituting approximately three-quarters of the recorded fauna as stated in Nel et al. (2013). Given the persistence of microbiological and insect species over time, their evolutionary dynamics could be thought to be one of the most efficient (adaptive), thus perhaps suggesting biological dominance. Human coexistence with these groups often acquires the characteristics of a defensive strategy due to their potential condition as vectors of diseases for human beings. Humans can 'dominate'—or populate with a certain level of efficiency—the macro world, explore the universe, and anticipate possible responses. Still, ultimately, they will not be able to conquer the microscopic world; if anything, yes, maybe control some of its menaces to the human species and establish mutually convenient partnerships.

What ancient biological inertias motorized human development and today underlie human behaviours and modern constructions, whether symbolic or virtual values, beliefs, dogmas, or norms? In this regard, the biological nature of the *Homo* construction involved millenniums of trials with variable rates of success. They included successful species during extended periods and others with comparatively short or unknown survival success. All of them provided traces of a variable and multiple evolutive chains. Changes in primaeval demographic distribution could have enabled the emergence or differential expansion of other lineages, generating transient stages in the evolutive display.

As discussed in Hubisz et al. (2020), gene flow would have occurred among various ancient hominin groups over the past several hundred thousand years. In this regard, several genetic admixture events would have taken place among hominins, including unknown archaic hominins in Africa and between unknown diverged or 'super-archaic' hominins (as *Homo erectus*), as well as more proximate ones such as among Neanderthals, Denisovans, and ancient humans. As quoted by Hubisz et al. (2020), perhaps the most well-studied example of archaic gene flow would be the interbreeding that occurred when humans migrated out of Africa and Neanderthals still roamed in Eurasia approximately 50,000 years ago:

> We identify 3% of the Neanderthal genome which is likely derived from gene flow from ancient humans. We also identify about 1% of the Denisovan genome that may be traced to an unsequenced archaic hominin; 15% of these regions were subsequently passed to modern humans. We find no convincing evidence that selection acted against any of these introgressed regions.

This gene flow would have occurred through introgression, i.e., incorporating novel genes and/or alleles from one taxon into the gene pool of a distinct taxon (usually via hybridization and backcrossing). Hammer et al. (2011) considered that polymorphisms in extant populations introgressed (i.e., gradual movement of genes from one species into the gene pool of another when there is some opportunity for hybridization) via relatively recent interbreeding with hominin forms that diverged from the ancestors of modern humans in the Lower-Middle Pleistocene. Though these authors considered that:

> Given recent fossil evidence, however, the greatest opportunity for introgression was in Africa, where AMH [anatomically modern humans] and various archaic forms coexisted for much longer than they did outside of Africa.

Renfrew (2006, 2008) referred to the *sapiens paradox* as the apparent uncoupling between genetic, physical, and cultural configurations of the *anatomically modern Homo sapiens.* Neurocognitive processes yielded the emergence of symbolic expression. On evolutive behavioural grounds, this is rooted in the temporal coincidence of parietal graphics and artefactual expressions and evidence of ritual behaviours, suggesting an interactive process with a brain physiologically apt to associate distributed complex neural circuits. In this regard, neural circuits or structures would have been related to specific functions and later recruited for other functional means—a process known as *exaptation*. This involves the recruitment of neural circuits associated with particular tasks for other functional means. It is frequently associated with new connectedness—*emergency*—whether implementing new or reinforced synaptic pathways.

On evolutive grounds, Ian Tattersall (2010, 2017) considered:

> The extraordinary cultural achievements of the Upper Paleolithic Europeans show that by 40 kyr ago the exploitation of the human symbolic potential was well advanced; but wherever and whenever this highly generalized capacity may have been initially discovered, substantially after the birth of Homo sapiens as an anatomical entity, the discovery of how to use it in new ways was a gradual process and one that is still proceeding.
>
> *(Tattersall, 2010)*

> Nothing indicates convincingly that any hominid behaved symbolically prior to Homo sapiens, which originated 200,000 years ago but started behaving symbolically only 100 k yr. later. Most probably the necessary neural underpinnings arose exaptively in the extensive developmental reorganization that gave rise to anatomically distinctive Homo sapiens and were recruited subsequently via a necessarily behavioral stimulus.
>
> *(Tattersall, 2017)*

As discussed in Colombo (2019), within these sets of proposals, the concept of *synaptic Darwinism* (Adams, 1998) applied to neural connections is cancelled or reinforced or generates conditional new ones, thus combining concepts of Darwin's theory applied to species evolution and of synaptogenesis, synaptolysis, and synaptic reinforcement at the microstructural level (Hebb, 1949).

Hammer et al. (2011) further considered that long-separated hominin groups exchanged genes with forms that either were evolving thoroughly modern features or had already attained a fully modern appearance, with some variants geographically placed in central Africa.

Mallick et al. (2016) discussed that admixture has been a dominant force in shaping patterns of genetic variation in human populations. According to Hubisz et al. (2020), a substantial proportion of the Neanderthal genome consists of regions introgressed from ancient humans, amounting to 3–7 per cent, suggesting that the Neanderthal genome would include a larger contribution from ancient humans than modern non-African human genomes. Scerri et al. (2018) argued that *Homo sapiens* evolved within a set of interlinked groups living across Africa, whose connectivity changed through time. Durvasula and Sankararaman (2020) stated:

> We provide complementary lines of evidence for archaic introgression into four West African populations. Our analyses of site frequency spectra indicate that these populations derive 2 to 19% of their genetic ancestry from an archaic population that diverged before the split of Neanderthals and modern humans.

In essence, an overview of the hominin emergence and gene flow among them and extant *Homo sapiens* provides a complex, interactive perspective of ancient *Homo* species connectivity and gene exchange before the emergence of *Homo sapiens,* as discussed in Kuhlwilm et al. (2016), Ackermann et al. (2019), Skov et al. (2018), and summarized by Stringer (2016):

> ... there are genetic traces in extant *H. sapiens* of earlier introgression from at least three extinct human groups. So how does this affect any definition of *H. sapiens* in the fossil record?

Sankararaman et al. (2016) addressed the reduction of Neanderthal ancestry among Europeans and East Asians, comparing similar introgression events among Denisovans, perhaps associated with a more significant load and purge of weakly deleterious alleles in Neanderthals linked to lower mutational fitness and smaller population. This would have involved *epistasis*—where the expression of one gene is modified (masked, inhibited, or suppressed) by the expression of one or more other genes, i.e., gene interactions that can affect phenotypes—, as discussed in Harris and Nielsen (2016) and Juric et al. (2016). Furthermore, a central notion should be mentioned placing human actions as a species within an evolutionary perspective. This is because the evolution of *Homo sapiens* occurred within what is configured as a species variant radiation, of which palaeontology is recovering some of its anatomical, functional, and molecular traces in attempting to build approximations to the biological identities of the different variants of the genus *Homo*. The data obtained from recent genetic studies converge in pointing out that hominids—before and after the emergence of anatomically modern *Homo*—exchanged genetic material. In 2006, Evans et al. claimed to have shown that haplogroup D was likely to have originated from a lineage separated from modern humans approximately 1.1 million years ago and introgressed into humans approximately 37,000 years ago. According to these authors, this finding supports the possibility of admixture between modern humans and archaic *Homo* populations (probably Neanderthals); through such an admixture, the human species would have benefitted evolutionarily by gaining new advantageous alleles. As these authors state, their results would provide genetic evidence in support of the possibility of admixture between modern humans and an archaic *Homo* lineage and additionally support the notion that the biological evolution of modern humans might have benefitted from the contribution of adaptive alleles from our archaic relatives.

Thus, humans would not have arisen from a genetically isolated line of close congeners from the phylogenetic point of view. Such is the case of the contribution of Neanderthals and Denisovans—considered *archaic Homo sapiens*—to the gene pool of the human genome (Harmon, 2012; Callaway, 2013). According to Pagani et al. (2016):

> Ancient DNA sequencing studies support an admixture between early Eurasians and at least two archaic human lineages, and suggest modern humans reached Eurasia at around 100 kya. Little additional archaic admixture, in Eurasia would have taken place over the last 35–45 thousand years.

Yet, more recently, Harris et al. (2023) considered that:

> ... results indicate that there have been multiple migration events of ancient modern humans (AMH) out of Africa and that Neanderthal and AMH gene flow has been bi-directional.

According to these authors, a group of extinct *Homo sapiens* from Africa would have interbred with Neanderthals in Eurasia around 250,000 years ago and left a genetic footprint in the DNA of Neanderthals that descended from this interbreeding event—with 6 per cent of the genome of a Neanderthal discovered in Siberia containing human DNA. Additionally, sub-Saharan populations of anatomically modern humans also inherited

Neanderthal DNA when groups of humans who had interbred with Neanderthals migrated back into Africa.

Thus, the current picture of the advent of later *Homo* species is ridden by admixtures with archaic *Homo* species—probably with different populations across the Eurasian continent before final speciation. Additionally, ancient genetic admixtures among related species complete a complex evolutive path of *Homo* radiation.

It seems necessary to mention that the current *Homo sapiens* evolutionary stage has not closed the evolution of *Homo*. Therefore, the evolutionary process continues, though now it does so from a species that is relatively conscious of its condition and that has instruments and technology that could start different paths or outcomes of this process. This represents a fundamental responsibility for the human species that poses ethical and social demands relative to the construction of our collective present and future.

Today, the anatomically modern *sapiens* species has managed to be the only surviving representative from the *Homo* radiation after overcoming the risks for survival in the natural kingdom. But it should be borne in mind that up to now, many of these other species—today's fossils—far exceeded the survival times humans have attained as a species. Additionally, material remnants indicate early hominin's constructive, creative, and technical cognition, based on the evidence of the earliest structural tool of wood for foraging and hunting (Barham et al., 2023). A differential characteristic concerning other *Homo* representatives—except for some species of 'archaic' *sapiens*, such as Neanderthals—has been the development of its brain volume and its complex, dynamic neurobiological characteristics, fundamentally due to the significant expansion of the cerebral cortex, expressed in part by the increased cranial capacity.

The relative increase in cerebral cortex grey matter—and its characteristics of neuronal density and connectivity—has been attributed to a possible relationship with the socialization process and its increasing complexity in primates, particularly in the genus *Homo* (Dunbar, 2003; Flinn et al., 2005; Roth and Dicke, 2005; Sallet et al., 2011). If we take this increase as an evolutionary parameter—limiting ourselves to hominids—, still corrected for body weight, this dimension generates a curve that goes from *Australopithecus* (about 3 million years ago, that is, before the emergence of *Homo*), with a cranial capacity of approximately 400 $cm^3$, to the 1,300 $cm^3$ (average) of the current human cranial capacity. However, evolution also incorporates brain structural and molecular connectivity changes involved in brain organization and functions that do not allow or exceed limited anthropometric comparisons with those of our fossil ancestors.

Unfortunately, reconstruction of the habits and capacities of the extinct *Homo* species is limited by the low number of or absence of material records of their possible behavioural profiles. The only traces that have remained are linked to their habitats, the production of instruments, the sharing of remains with other specimens, traces in the composition of the soil and possible eating habits, body dimorphisms of gender, and comparisons with surviving species of primates. Some positive evolutionary components—adaptive evolution or positive selection—would have been involved in such brain development, to the extent that the human brain continues its growth for a long time after birth, and its essential organization, fundamentally regarding cortical neural circuits (Coqueugniot et al., 2004;

De Silva and Lesnik, 2008), is not completed until adolescence. This implies a significant degree of initial neurobiological immaturity and helplessness, which must have required rearing efforts from the human species that were longer than in other species. Such a handicap to completing this postnatal period suggests the emergence of an evolutionarily favourable combination of biological and social factors.

The cerebral energy consumption during the early postnatal period—close to 80 per cent of the total body in a resting state—indicates the degree of such demand and of the significant fragility of this period. It also suggests the possibility of an increased modulatory/permissive impact of early emotional/social/cultural factors on developmental processes. During the germinal times of the species, these demands would have required an influential community organization in the face of natural risks imposed on child survival.

The cerebral cortex's extended developmental period, which involves cell production, cell migration, synaptic connectivity with subsequent pruning, and its intracortical and subcortical connectivity, makes it labile to environmental agents and rearing conditions. Its functional consequences on brain connectivity and cognitive and emotional development are placed at stake under conditions involving poverty, indigence, and marginality that affect early mental development (Colombo 2007; Lipina and Colombo 2009).

A crucial step in human evolution has been transitioning from prey and predator during the gatherer/scavenger/hunter stages to controlling the Earth's energy yield. This allowed humans to concentrate on controlling their environment and competing with the threats imposed by their potential and formidable rivals: their congeners, possessors of similar capacities.

The shift from prey to universal predator affected the organization of the human brain, its behaviour, and social organization. But the human species also had to deal with a rival that emerged from their own mental development and own conscience: the intimate notion of death, of existential finiteness, and the need to control the resulting emotional impact using all the resources of human imagination and rationality. This progressively compromised or conditioned their freedom differently from the physical threat of competitive fellow species.

After centuries and millennia, the 'Darwinian jungle' has changed its actors, but the essence persists. For what reasons did the other *Homo* species become extinct? In general, the vectors for survival involve the adaptive capacity of the genetic menu, the demands imposed by the environment—climate, predators—and the nutritional and food requirements and availability of the species, as well as interspecies competition. In short, its survival would be represented by the result of an equation between two terms: the efficiency of the species for its individual subsistence and vegetative growth and the complex demands imposed by the environment. Even within the genus *Homo*, the radiation of species—whose actual number is probably still incomplete—does not build a metaphorical pyramid but a fan, from which *Homo sapiens* managed to survive after coexisting for some millennia with four other *Homo* species. From poachers, hunter-gatherers, and scavengers, humans progressively began to control the energy yield and biodiversity of the Earth, and with it came comparatively improved conditions for settlement or territorial anlage, the progressive demographic explosion, and coarse, unstable social organization. Without realizing it, humans began to compromise the ecological footprint (a measure of the demand imposed on the production of the Earth's ecosystems), that is, the environmental limits of the planet to our food, productivity, and consumption demands.

Approximately 15 thousand years ago, *Homo sapiens* would have become the only survivor of the *Homo* species. However, traces of previous *Homo* species have been imprinted in their genetic code and have contributed to the variable genetic and phenotypic construction of functional and anatomically modern *sapiens.* No other species then lived up to its competence, to its Machiavellian intelligence. This forced human communities to reconsider survival and dominance strategies when facing new competitors. Members of the *sapiens* species began an internal competition for pre-eminence and dominance in the natural kingdom. Eventually, differences throughout the history of civilization regarding the destruction, assimilation, or enslavement of its opponents, together with bids for interactive access to scientific-technological achievements of different cultures, contributed to further fractured human development and sources of conflicts. Concomitant with these conditionings arose the projections of the animist world that were progressively reformulated in dogmatic mystical proposals and religious policies that reprogrammed human interactions under different profiles of dominance.

However, the feeling of power over natural resources and the awareness of human creative capacity faced the challenge emerging from those of its peers. It also did so with a rival that emerged from its conscience of finiteness and survival drive.

## Notes

1 When two species coexist in the same geographic area; sympatric speciation refers to the process through which new species evolve from an ancestral one, while they share the same territory.

## References

Ackermann RR, Arnold ML, Baiz MD, Cahill JA, et al. 2019. Hybridization in human evolution: Insights from other organisms. *Evol Anthropol* 28: 189–209. doi: 10.1002/evan.21787.

Adams PJ. 1998. Hebb and Darwin. *J Theor Biol* 195: 419–438.

Barham L, Duller GAT, Candy I, Scott C, et al. 2023. Evidence for the earliest structural use of wood at least 476,000 years ago. *Nature* 622: 107–111. https://doi.org/10.1038/s41586-023-06557-9

Callaway E. March 2013. Most popular human cell in science gets sequenced. *Nature News.* doi:10.1038/nature.14196.

Callaway E. April 2021. Oldest DNA from a *Homo sapiens* reveals surprisingly recent Neanderthal ancestry. *Nature News.*

Colombo JA. 2007. Acerca del desarrollo cerebral infantil (On brain child development). In JA Colombo (editor), *Pobreza y desarrollo infantil* (*Poverty and child development*). Paidos Editorial.

Colombo JA. 2019. *Our Animal condition and social construction.* Nova Science Publishers. ISBN: 978-1-53615-357-6. eBook ISBN: 978-1-53615-3583.

Colombo JA. 2022. *Dominance behavior: An evolutive and comparative perspective.* Springer Nature.

Coqueugniot H, Hublin J-J, Veillon F, Houet F, Jacob T. 2004. Early brain growth in Homo erectus and implications for cognitive ability. *Nature* 431: 299–302.

De Silva JM, Lesnik JJ. 2008. Brain size at birth throughout human evolution: A new method for estimating neonatal brain size in hominins. *J Human Evol* 55: 1064–1074.

Devièse T, Karavanić I, Comeskey D, Kubiak C, et al. 2017. Direct dating of Neanderthal remains from the site of Vindija Cave and implications for the Middle to Upper Paleolithic transition. *Proc Natl Acad Sci USA*. 114: 10606–10611. https://doi.org/10.1073/pnas.1709235114.

Dunbar RIM. 2003. The social brain: Mind, language, and society in evolutionary perspective. *Ann Rev Anthropol* 32: 163–181.

Durvasula A, Sankararaman S. 2020. Recovering signals of ghost archaic introgression in African populations. *Science Advances* 6(7). doi: 10.1126/sciadv.aax5097.

Evans PD, Mekel-Bobrov N, Vallender EJ, Hudson RR, Lahn BT. 2006. Evidence that the adaptive allele of the brain size gene *microcephalin* introgressed into *Homo sapiens* from an archaic *Homo* lineage. *Proc Natl Acad Sci USA*. 103(48): 18178–18183. doi: 10.1073/pnas.0606966103.

Flinn MV, Geary DC, Ward CV. 2005. Ecological dominance, social competition, and coalitionary arms races: Why humans evolved extraordinary intelligence. *Evol Hum Behav* 26: 10–46.

Foley NM, Springer MS, Teeling EC. 2016. Mammal madness: Is the mammal tree of life not yet resolved? *Philos Trans R Soc B* 371: 20150140. http://dx.doi.org/10.1098/rstb.2015.0140.

Hammer MF, Woerner AE, Mendez FL, Watkins JC, Wall JD. 2011. Genetic evidence for archaic admixture in Africa. *Proc Natl Acad Sci* 108: 15123–15128.

Harmon K. 2012. New DNA analysis shows ancient humans interbred with Denisovans. *Nature*. doi 10.1038/nature.2012.11331.

Harris DN, Platt A, Hansen MEB, Fan S, et al. 2023. Diverse African genomes reveal selection on ancient modern human introgressions in Neanderthals. *Curr Biol* 33(22): 4905–4916. https://doi.org/10.1016/j.cub.2023.09.066.

Harris K, Nielsen R. 2016. The genetic cost of Neanderthal introgression. *Genetics* 203: 881–891. doi: 10.1534/genetics.116.186890.

Hebb DO. 1949. *The organization of behavior. A neuropsychological theory*. Wiley.

Hubisz MJ, Williams AL, Siepel A. 2020. Mapping gene flow between ancient hominins through demography-aware inference of the ancestral recombination graph. *PLoS Genet* 16(8): e1008895, 2020. https://doi.org/10.1371/journal.pgen.1008895.

Hublin J-J. 2017. The last Neanderthal. *Proc Natl Acad Sci USA* 114: 10520–10522. doi: 10.1073/pnas.1714533114.

Juric I, Aeschbacher S, Coop G. 2016. The strength of selection against Neanderthal introgression. *PLoS Genet* 12(11): e1006340. doi: 10.1371/journal.pgen.1006340.

Kuhlwilm M, Gronau I, Hubisz MJ, Filippo C, et al. 2016. Ancient gene flow from early modern humans into Eastern Neanderthals. *Nature* 530: 429–433. doi: 10.1038/nature16544.

Lipina SJ, Colombo JA. 2009. *Poverty and brain development during childhood*. American Psychological Association.

Mallick S, Li H, Lipson M, Mathieson I, et al. 2016. The Simons Genome Diversity Project: 300 genomes from 142 diverse populations. *Nature* 538: 201–206.

Mora C, Tittensor DP, Adl S, Simpson AG, Worm B. 2011. How many species are there on Earth and in the ocean? *PloS Biology* 9(8): e1001127.

Nel A, Roques P, Nel P, Prokin AA, et al. 2013. The earliest known holometabolous insects. *Nature* 503: 257–261. doi: 10.1038/nature12629.Epub.

Pagani L, Lawson DJ, Jagoda E, Mörseburg A, et al. 2016. Genomic analyses inform on migration events during the peopling of Eurasia. *Nature* 538: 238–242. doi: 10.108/nature19792.

Renfrew C. 2006. Becoming human: The archaeological challenge. *Proc Br Acad* 139: 217–238.

Renfrew C. 2008. Neuroscience, evolution and the sapient paradox: The factuality of value and of the sacred. *Philos Trans R Soc* B 363: 2041–2047. doi:10.1098/rstb.2008.0010.

Roth G, Dicke U. 2005. Evolution of the brain and intelligence *Trends Cogn Sci* 9: 250–257.

Sallet J, Mars RB, Noonan MP, Andersson JL, et al. 2011. Social network size affects neural circuits in macaques. *Science* 334: 697–700.

Sankararaman S, Mallick S, Patterson N, Reich D. 2016. The combined landscape of Denisovan and Neanderthal ancestry in present-day humans. *Curr Biol* 26: 1241–1247. doi: 10.1016/j.cub.2016.03.037.

Scerri EML, Thomas MG, Manica A, Gunz P, et al. 2018. Did human species evolve in subdivided populations across Africa, and why does it matter? *Trends Ecol Evol* 33: 582–594. doi: 10.1016/j.tree.2018.05.005.

Skov L, Hui R, Shchur V, Hobolth A, et al. 2018. Detecting archaic introgression using an unadmixed outgroup. *PLoS Genet* 14(9): e1007641. https://doi.org/10.1371/journal.pgen.1007641.

Stringer C. 2016. The origin and evolution of *Homo sapiens. Philos Trans R Soc Lond B Biol Sci* 371(1698): 20150237. doi: 10.1098/rstb.2015.0237.

Tattersall I. 2010. Human evolution and cognition. *Theory on Biosciences* 129: 193–201. doi: 10.1007/s12064-010-0093-9.

Tattersall, I. 2017. How can we detect when language emerged? *Psychon Bull Rev* 24: 64–67. https://doi.org/10.3758/s13423-016-1075-9.

Vignieri S. 2023. Zoonomia. *Science* 380: 356–357. doi: 10.1126/science.adi1599.

# 3

# FURTHER INSIGHTS ON *HOMO SAPIENS'* EVOLUTION

Within the large and complex variety of living forms, the evolution of *Homo sapiens* emerged within what configures a radiation—as opposed to a pyramid—for which paleoanthropology studies are attempting to recover some of the anatomical, functional, and molecular traces to build a scientific approach to the biological identities of different variants of the genus *Homo*. The data obtained from recent genetic studies would converge in pointing out that—before and after the emergence of anatomically modern man—the hominids exchanged genetic material. Thus, from the phylogenetic point of view, *Homo sapiens* would not have arisen from an isolated genetical line. Such would be the case for the contribution of the Neanderthals and Denisovans—considered archaic *Homo sapiens*—to the genetic pool of the human genome, as proposed by Harmon (2012) and Callaway (2013). Thus, the emergence and evolution of *Homo sapiens* would not be germane to ancient evolutive drives and their social complexity.

For at least 15 thousand years, the species *Homo sapiens* managed to survive its direct competitors and to maximize obtaining resources from the planet. At that time, humans were not aware of the dimensions of that competition. This only happened when they became aware of the species spread around the planet and the increased demand on the live and physical environment. In the beginning, humans shared the risks and advantages of being one species *Homo*, in the balance of biodiversity. With time, humans distorted their relationship with the environment until it affected all ecosystems and accelerated the extinction of animal and vegetal species or increased the survival risk of others that competed for similar ecological niches and resources.

This continued until the human species became aware that its powerful enemy was of comparable strength. The opponent was formidable, so successively, humans developed increasingly powerful and ominous instruments of destruction and defence, for dominant drives did not abandon the new *Homo* species.

Within the complex variety of biological species there arose a manifest relativism in the human species' supposed status of 'dominant species', for humans are in a broad,

DOI: 10.4324/9781003571544-4

interactive niche limited by niches of the microworld in which other species are the dominant ones. These include microorganisms that share human ecological contexts, and those that survive in harsher, more extreme conditions for 'life' as we know it. These are the so-called 'extremophile' microorganisms.

According to Kaessman and Pääbo (2002), besides the basic neural organization and interactions, the nucleotides that have been passed down to *Homo sapiens* from their ancestors in every generation were modified by mutations. These may accumulate in characteristic ways and generate slightly different versions of the ancestral genome that were passed on to subsequent generations. Yet, embrained in core neural circuits (mostly in the basal brain), survive our basal drives (reproductive, territorial, survival, feeding), basic responses (fight, flight), and the thresholds for their behavioural expression.

*Homo sapiens* would have shared a common ancestor with earlier primates, represented 5–6 million years ago by the chimpanzee and bonobo, with the gorilla 7–8 million years ago, and with the orangutan 12–13 million years ago (Gagneux and Varki, 2001). This primate evolution indicates uncertainties regarding genetic species divergence, as discussed by Varki and Nelson (2007). Based on the genetic proximity of the chimpanzee to the human genome, these authors suggest modifications to taxonomic concepts of the genre *Homo*. However, current taxonomic classification places the chimpanzee as a *hominin*. According to Curnoe and Thorne (2003) and Chaimanee et al. (2012), chimpanzees share the same genre *Homo* with humans.

According to Osada and Wu (2001) and Patterson et al. (2006), *Homo sapiens* and chimpanzees would have diverged less than 6.3 million years ago. Additionally, these authors considered the possibility of genetic exchange between them before attaining final speciation. As stated by Kehrer-Sawatzki and Cooper (2007), DNA nucleotide differences between them would range to approximately four per cent. According to Foley et al. (2016) hominin evolution was the product of a complex interaction with variable rates of genetic exchange depending on environmental dynamics and competitive interactions with other contemporary hominins.

As stated by Preston (2013):

> ... even complex human behaviors reflect ancient mammalian neural systems that evolved to solve key problems in adaptive ways, with far reaching consequences for even our most venerated human traits.

More recently, Friederici et al. (2024) proposed:

> A key to uncovering mechanisms of hominid brain evolution is understanding the outstanding plasticity of primate brains (Sherwood and Gómez-Robles, 2017; Froudist-Walsh et al., 2018). Environmental factors shape the brain and the behavior, resulting in environmental variation across individuals and populations (Logan et al., 2018).
>
> Understanding brain evolution, therefore, requires examining brains from individuals that have lived in variable socioecological environments, importantly, including the

> species' natural habitat (Bruner, 2019). Studies conducted only on animals from similar socioecological environments, for example from research centers, may limit ...
> and systematically bias our knowledge on great ape neuroanatomy and brain plasticity.

These concepts provide an evolutive foundation for ancient neurobehavioural drives anchored in human behaviour. They include possible ancient interactions and genetic exchanges among several hominids and hominin lineages in human genetic composition, which underlies our diverse, sophisticated, complex current human culture and ethnic diversity. Thus, humans are not pure breed.

Thus, the evolution of *Homo sapiens* would not be foreign to ancient evolutive drives and social complexity, as proposed by Dunbar (2009) and Dunbar and Shultz (2007). Further complementary proposals include the concept of a shift in feeding behaviour with the subsequent reduction of the digestive system to compensate for energy demand due to brain growth (Aiello and Wheeler, 1995) (though questioned by Navarrete et al., 2011, as discussed later); and genetic modifications and neurotransmitter involvement, such as interactions of dopamine and cultural norms (Previc, 1999, 2009). These would have evolved at different times during the evolution of *sapiens*.

This proposed multifactorial composite should include the potential role of the *gut microbiome* that would represent an additional physiological factor (defining a multigenic organism involving host endogenous DNA and a universe of coexisting bacteria associated with feeding habits) linked to existing differences among primate species and projecting onto brain and behavioural domains, as discussed in Colombo (2024a). Dietary and phylogenetic factors (Ley et al., 2008) would regulate gut microbiome population and composition, which generates a species diversity schematically represented by carnivores, omnivores (as humans are), and herbivores. This involves consequences for the structure, physiology, and development of the nervous system and behavioural profiles based on nutritional and metabolic interactions that—at the human level—ought to be included in the construction of public policies applied to child nutrition.

DNA traces of fossils remaining in the soil provide a basis for research on the construction of the probable physiological profile of our early ancestors, as analysed by Slon et al. (2017) in sediments from seven archaeological sites covering a period of 550,000-14,000 years. As summarized in Colombo (2019):

> The biological nature of our construction drags millenniums of trials with variable rates of success: of successful species during extended periods and others with comparatively short or unknown survival success. All of them provided traces of a variable and multiple evolutive chains.

*Homo erectus* represents a long-lived species that would have made several migrations within Africa and led to migrations towards the Euro Asiatic continent. *Homo erectus* or *Homo* variations that spanned from it (e.g., *Homo antecessor*, *Homo heidelbergensis*) would have eventually led to the emergence of *Homo sapiens*. Enclaves left in such an extended path would have been later invaded in different regions by other migrant *Homo* species, including the *anatomically modern sapiens* and Neanderthals. These migrations would have led to interbreeding, giving rise to ecological adaptations and generating different phenotypes and ethnicities. An editorial commentary published by *Nature* (2016)[1] reported on the vast variety of hominids that preceded the human species, some of them occasionally cohabiting with the modern *sapiens*, resulting in a large collective that would have led to interbreeding and had an impact on genetic mixing and human species evolution.

Foley et al. (2016) and Lahr (2016) attempted to stage the evolutive process of the hominids, with emphasis on the transitions resulting from reductions in the demographic density of various lineages. Conditions affecting their primaeval demographic distribution would have provided conditions for the emergence or differential expansion of lineages representing transient stages in the evolutionary process.

Following considerations by Kuhlwilm et al. (2016), the hypothetical evolutive chain leading to the human species would recognize different routes and possible genetic exchanges among intervening *Homo* evolutive stages. Consequently, it could be concluded that our closest ancestors would have evolved from the interaction of a variety of lineages following a succession of hominid species, each one of them characterized by structural components and functional and survival strategies, as proposed by Foley et al. (2016), Kuhlwilm et al. (2016) and Lahr (2016).

Epigenetic factors would have completed the speciation of the modern human and conditioned the probable functional profiles of our ancestors within a dynamic context, including basic behaviours that involved territorialism, sexual drive, individual and community survival drives, and securing feeding resources. These dynamic processes pose the probability of friction, as expressed metaphorically, between the neurobiological and cultural *tectonic plates*, as mentioned in Colombo (2010, 2013, 2015).

Though brain/mental plasticity allows for adaptative responses on behavioural grounds, it operates on a basic framework of primary drives imprinted in the core of our animal construction and modulated by cultural contexts. In such regard, psychological mechanisms would disguise or temper some of the behavioural consequences implied in such frictions or uncoupling processes involving the described basic tectonic plates.

The relationships between individuals, the group structure to which they belong, and the environment profile reformulate the expression of the neurobiological substrate that conditions individual behaviour in gregarious species. The increasingly complex social demands placed on the neurobiological organization in complex social structures configure an interactive universe typical for each species, which acquires greater complexity in species of the genre *Homo* as described in Colombo (2024a, 2024b). This implies an additional demand on brain/mind processing to the basic behavioural profiles (dominance, survival, gregariousness) present in every human and non-human gregarious animal community with a hierarchical organization.

In the late *Homo* species, brain internalization of culture must have been a slow process due to its evolving and implicit complexity, progressively based on the interaction of values, beliefs, language, and symbolic constructs, coupled with other survival conditions. In *Homo sapiens*—taking advantage of the long period of postnatal maturation—emerged the diversity of 'social phenotypes' that are glimpsed as the collective substrate of the different nations or ethnic groups. This interactive, dynamic process would have stimulated brain microstructural changes in connectedness, affecting neural circuits associated with a social organization that modulated the expression of the phenotypic behaviour.

Boyd and Richerson (2009) and Kitayama and Uskul (2011) stressed the role of biocultural interactions as an environmental configuration factor. The culturally evolved environments affect which genes are favoured by natural selection, while genetic characteristics affect the expression of certain forms of culture, as stated by Kitayama and Uskul (2011). In this biosocial interaction, the brain/mind acts as the substrate of such biosocial acting vectors implying the genetic inheritance expressed in brain organization, as well as the generator of cultural events that modulate such expression. These mutual conditionings of individual and collective phenotypes have been operating for centuries, with structural complexities depending on the interactions within gregarious groups and their socioecological environment. Basic behavioural configurations are usually crystallized by the repetition of individual and collective behavioural patterns, to which humans added complex brain/mind capabilities, including the emergence of a language that fostered virtual interoceptive and exteroceptive constructions. Thus, the degrees of freedom by which an individual relates to the physical and emotional environmental demands were expanded, though remaining conditioned by early neurobehavioural imprints that more fully emerge under critical conditions.

All animal gregarious species express behavioural modes adapted to each species in what could be considered protocultural expressions. In *Homo sapiens*, the nature and acquired behavioural complexity conditioned by language development would have required a long-term evolutive and ontogenetic process. On the grounds of relative time consideration, according to some authors, such cultural development would have had either a long, slow, progressive format or an accelerated one—the so-called *cultural revolution*—during the last 40,000 years (as discussed in Renfrew, 2008 and Renfrew et al., 2008), linked to the emergence of language. According to Mcbrearty and Brooks (2000), the emergence of *Homo sapiens* in its strict sense has been tentatively dated more than 100,000 years ago. Considering that the claimed emergence of modern human behaviours known as the *human revolution* would have occurred ca. 40,000–50,000 years ago, the earliest modern human African individuals would have expressed behaviourally primitive profiles. Following Colin Renfrew (2006, 2008), this uncoupling between the genetic, physical, and cultural configuration of the *anatomically modern Homo sapiens* is represented by what has been called the *sapiens paradox*. Considering material evidence implying the temporal coincidence of parietal graphics and artefactual expressions and evidence of ritual behaviours, progressive neurocognitive modifications must have taken place to develop capabilities for symbolic expression in a behavioural interactive context. This would suggest an interactive process with a physiologically evolving brain progressively able to associate distributed neural circuits. In such a process, phenomena of *exaptation*—recruitment of neural circuits associated with specific functions for other functional means—and of *emergence* would have occurred when new connectivity was involved.

In the genetic domain, the culturally modified physical and interactive conditions affected the natural selection of dominant genes. This, in turn, would have affected interpersonal and material interactive behavioural domains, as suggested by Kitayama and Uskul (2011). In this biosocial interaction, the brain/mind is a depositary of both acting vectors (genetic and environmental). These vectors have interacted through centuries as conditioners of individual and collective behavioural phenotypes. The repetition of individual and collective behavioural patterns would have led to the crystallization of basic behavioural configurations (dominance, fight/flight behaviour, and survival priorities) with species-specific characteristics. Additionally, communication codes allowed for increased accuracy and prediction of behavioural responses, thus expanding the *degrees of freedom* with which an individual relates to the physical and emotional environment.

The differential cultural slope among evolving human communities was affected by continued, disparate technological development that permeated most cultural niches, although with different cultural goals and results. This eventually resulted in a technological brain incarnation with different applications depending on the cultural profiles of the different human communities. This is to the point that technological and instrumental applications significantly modulate human priorities and behavioural profiles, in several instances approaching a state of technocracy and affecting dominance patterns. The technological impact would have occurred when the first notion of 'instrument' entered our socioecological niche, with an increasing slope of its power to mould behaviour, currently affecting our very structure of values and cultural scaffolding.

Should we place into perspective previous comments on biosocial interactions (Cole ER, 2009a) in several human communities, the progressive encephalization of a technocratic culture and the gradual birth of a post-human civilization? This has primarily anchored in certain community strata; thus, it is expected that this *embrainment* of *technological culture* has its impact as an epigenetic factor and would differentially affect genetic expression.

This process will tend to deepen the current segmentation process of the human community, further contradicting the general concept of a globalized world.

## 'Globalized' or 'Segmented'?

Unfortunately, opinion makers have synthesized our world by coining the term *globalization* to define a more vital link between economic events and information. This term has replaced the use of *humanity* and, in doing so, has redefined the interactions of the world community irrespective of any human inclusion. The term globalization is anomic; it ignores the individual.

It is argued that globalization is an economic, technological, social, and cultural process on a grand scale, which increased communication and interdependence among countries worldwide by unifying their markets, societies, and cultures. Yet, this would seem to neglect that—though with diverse impact on human behavioural configuration—such an intercommunication trend has been in progress since humans invaded all planetary niches. Their respective populations grew, and more yet, its increasing information sharing and interactions would have probably been one central cause of the rapid evolution of human civilization. Conquest, colonization, artistic and literary influences, and dominance have long been among us, whether to achieve greater revenues or energy resources or to satisfy expansionist tendencies—whether territorial, cultural, or financial. But, even more important, such a concept seems to disregard the asymmetries that profoundly divide the living conditions of communities and of individual citizens within them. Thus, the world does not operate as an *equal opportunity* employer, nor an *equal opportunity* educator, nor does it provide conditions for an *equal opportunity* citizenship.

Let us recognize the importance of the speed of information used nowadays in the decision-making process. However, we should also realize that spy satellites and the non-proliferation of nuclear weapons are issues solely controlled by a small group of superpowers that supposedly struggle to maintain a feeble 'peace' condition in the world or protect their own interests and prevailing conditions. Around them revolve the rest of the world's communities, aware of the inequity but often unable or afraid to initiate a change. The thesis is that the so-called *globalization* runs the risk of deepening the *segmentation of humanity* in terms of rates of change, quality of life, dominance, and educational level. When the rate of social, scientific, and technological change is not uniform but rather significantly asymmetric, what tends to happen is that, as in any competitive environment, dissimilar groups begin to form. The front-runners set the pace. The rest are followers, or are excluded and become marginals or dropouts, or decide to opt for other directions, which may develop into conflict.

As discussed in SW Cole (2009b), biosocial interactions provided grounds for the progressive encephalization of technological culture and the emergence of what has been defined as a *post-human civilization*. This involved the emergence of both agriculturally and technologically based social profiles with different rates of social dynamics and changes. This asymmetrical development questions the real meaning of the publicized concept of *globalization*, as opposed to the concept of *global balkanization* based on a grand divide between scientifically/technologically advanced societies—as exemplified in the proposals involving the advantages of mineral exploitation in meteors and planets (Elvis, 2013)—and those societies that are not advanced in that way, and the influence of profit and marketing in setting up sociocultural profiles. The ancient drive of setting up individual and collective standings based on dominance continues to influence contemporary social behaviours. Uneven technological and scientific developmental cultural driving profiles contribute to the further development of a conflictive humanity and the disparity of dominance profiles.

If behaviour driven by dominance and profit completely replace concepts based on sharing, solidarity, and collective opportunities for development, the above considerations place a worrying perspective on human development. There seems to be a profound need to change cultural gear on the motivation for the political use and promotion of technological progress. Additionally, socioeconomic segmentation in the access to knowledge, informed behaviour, and creativity places a new impetus on the balkanization process of the human community. Since access to valued information is restricted/conditioned for economic, sociocultural, and political reasons, it generates strata of access to information in addition to an educated, critical access to it. Such interactive domains eventually set up the stage and lead towards the promotion of dominant behaviours, where they could acquire the role of oppressor or the generator of social prejudices.

## Note

1 *Nature Editorial* (2016) https://www.nature.com/articles/d41586-01806657-x?

## References

Aiello LC, Wheeler P. 1995. The expensive-tissue hypothesis—The brain and the digestive system in human and primate evolution. *Current Anthropology* 36: 199–221.

Boyd R, Richerson PJ. 2009. Culture and the evolution of human cooperation *Philos Trans R Soc Lond B Biol Sci* 364: 3281–3288.

Callaway E. 2013. Most popular human cell in science gets sequenced. *Nature News*. doi:10.1038/nature.14196.

Chaimanee Y, Chavasseau O, Beard KC, Kyaw AA, et al. 2012. Late Middle Eocene primate from Myanmar and the initial anthropoid colonization of Africa. *Proc Natl Acad Sci USA* 109: 10293–10297.

Cole ER. 2009a. Intersectionality and research in psychology. *Am Psychol* 64: 170–180. https://doi.org/10.1037/a0014564.

Cole SW. 2009b. Social regulation of human gene expression. *Curr Dir Psychol Sci* 18: 132–137. doi: 10.1111/j.1467-8721.2009b.01623.x.

Colombo JA. 2010. *Somos la especie equivocada? (Are humans the mistaken species?)*, Eudeba. https://tinyurl.com/lsdozicg.

Colombo, JA. 2013. *Bajo libertad condicionada* (*Under conditioned freedom*). Imago Mundi.

Colombo JA. 2015. *Los Homo sabios. Globalizados o segmentados?* (*The wise Homo. Globalized or segmented?*) Buenos Aires Books.

Colombo JA. 2019. *Our animal condition and social construction*. Nova Science Publishers. ISBN: 978-1-53615-357-6. eBook ISBN: 978-1-53615-3583.

Colombo JA. 2024a. *Evolution and the human-animal drive to conflict. A psychobiological perspective*. Routledge.

Colombo JA. 2024b. *Human biome and human behaviour*. Routledge.

Curnoe D, Thorne A. 2003. Number of ancestral human species: A molecular perspective. *Homo* 53: 201–224.

Dunbar RIM. 2009. The social brain hypothesis and its implications for social evolution. *Ann Hum Biol* 36: 562–572. doi: 10.1080/03014460902960289.

Dunbar RIM, Shultz S. 2007. Evolution in the social brain. *Science* 317: 1344–1347. doi: 10.1126/science.1145463.

Elvis M. 2013. Prospecting asteroid resources. In V Badescu (editor), *Asteroids*. Springer. https://doi.org/10.1007/978-3-642-39244-3_4.

Foley R, Martin L, Lahr MM, Stringer C. 2016. Major transitions in human evolution. *Philos Trans R Soc London B* 371: 20150229. http://dx.doi.org/10.1098/rstb.2015.0229.

Friederici AD, Wittig RM, Anwander A, Eichner C, et al. 2024. Brain structure and function: A multidisciplinary pipeline to study hominoid brain evolution. *Front Integr Neurosci* 17: 1299087. doi: 10.3389/fnint.2023.1299087

Gagneux P, Varki A. 2001. Genetic differences between humans and great apes. *Mol Phylogenetics Evol* 18: 2–13.

Harmon K. 2012. New DNA analysis shows ancient humans interbred with Denisovans. *Nature*. doi: 10.1038/nature.11331.

Kaessmann H, Pääbo S. 2002. The genetical history of humans and the great apes. *J Intern Med* 251: 1–18.

Kehrer-Sawatzki H, Cooper DN. 2007. Understanding the recent evolution of the human genome: Insights from human-chimpanzee genome comparisons. *Hum Mutat* 28: 99130.

Kitayama S, Uskul AK. 2011. Culture, mind, and the brain: Current evidence and future directions. *Annu RevPsychol* 62: 419–449.

Kuhlwilm M, Gronau I, Hubisz MJ, Filippo C, et al. 2016. Ancient gene flow from early modern humans into Eastern Neanderthals. *Nature* 530: 429–433. doi: 10.1038/nature16544.

Lahr MM. 2016. The shaping of human diversity: Filters, boundaries and transitions. *Philos Trans R Soc B* 371: 1–12. http://dx.doi.org/10.1098/rstb.2015.0241.

Ley RE, Hamady M, Lozupone C, Turnbaugh P, et al. 2008. Evolution of mammals and their gut microbes. *Science* 320: 1647–1651. doi: 10.1126/science.1155725

Mcbrearty S, Brooks AS. 2000. The revolution that wasn't: A new interpretation of the origin of modern human behavior. *J Hum Evol* 39: 453–563.

Navarrete A, van Schaik CP, Isler K. 2011. Energetics and the evolution of human brain size. *Nature* 480: 91–93.

Osada N, Wu CI. 2001. Inferring the mode of speciation from genomic data. *Genetics* 169: 259–264.

Patterson N, Price AL, Reich D. 2006. Population structure and Eigen analysis. *PLoS Genet* 2(12): e190. https://doi.org/10.1371/journal.pgen.0020190.

Preston SD. 2013. The origins of altruism in offspring care. *Psychol Bull*. doi: 10.1037/a0031755.

Previc FH. 1999. Dopamine and the origins of human intelligence. *Brain and Cognition* 41: 299–350.

Previc FH. 2009. *The dopaminergic mind in human evolution and history*. Cambridge University Press. https://doi.org/10.1017/CBO9780511581366.

Renfrew C. 2006. Becoming human: The archaeological challenge. *Proc Br Acad* 139: 217–238.

Renfrew C. 2008. Neuroscience, evolution and the sapient paradox: The factuality of value and of the sacred. *Philos Trans R Soc B* 363: 2041–2047. doi:10.1098/rstb.2008.0010.

Renfrew C, Frith CD, Malafouris L. 2008. Introduction. The sapient mind: Archaeology meets neuroscience. *Philos Trans R Soc B Biol Sci* 363: 1935–1938.

Slon V, Hopfe C, Weiss CL, Mafessoni F, et al. 2017. Neandertal and Denisovan DNA from Pleistocene sediments. *Science* 356: 605–608. doi: 10.1126/science.aam9695.

Varki A, Nelson DL. 2007. Genomic comparisons of humans and chimpanzees. *Annual Review on Anthropology* 36: 191–209.

# 4

# BIOLOGICAL NATURE AND CULTURAL CONSTRUCTION

## The Concept of Tectonic Plates

The accelerated progress in scientific and technological domains rides on our ancestral animal construction of behavioural drives, instincts, and basic needs. Three events allowed the construction of a culture that attempts to harmonize innate, primal components and drives—'needs of infrastructure construction'—lodged within the human animal behavioural 'core'. First, were the emergence and development of successive brain neural circuits associated with a neocortex whose evolutive complexity and plasticity allow continuous reprogramming of human representations of reality and self. Second, this neurodevelopmental process is associated with events involving the inhibitory processing of 'primary impulses' during the postnatal extension period of brain organization. Third, this made possible the social carving of brain/mental organization and human social complex behaviours, and the emergence of rich language communication. These allowed the species to build representations, abstractions, social norms, and projections of material and imagined 'reality'.

Even so, as mentioned earlier, these last *tectonics*—a metaphorical description—of the human cultural niche are mounted on an additional one that preceded them: the 'neurobehavioural tectonic plate' built on drives of primal animal origin.

In general, the vectors for the survival of species are built on the genetic menu, on the adaptive range of ability to meet the demands imposed by the environment—climate, predators—and on the nutritional and food requirements of the species. The human brain and behavioural adaptations emerged from those composite evolutive conditions until they grew to adapt and actively condition the range of environmental events.

Yet, the species that took eons to become an additional influential character in natural history has managed to create in about 5,000 years expressions of experiences that can be transmitted; barely 200 years ago, it managed to use technology to replace man in many tasks; and from about 70 years ago it began to scratch the mysteries of outer and subatomic space. Yet, the fast growth in scientific-technological advances should not create a false image that the primal memory of our basic animal construction has been erased

DOI: 10.4324/9781003571544-5

from human basal neural circuits' heritage, for humans did not cease dealing with the expression and management of biological drives, 'instincts', and essential survival needs to be embedded in an emotional dimension. These find their expression in social interactions and critical survival instances, where dominance, oppression, and physical and creative destruction reveal the continuous background presence of ancient animal drives.

As mentioned above, interactive events allowed the construction of a culture that attempts to cope with those innate tendencies, involving the emergence of a neocortex, the extended postnatal period of cerebral organization, and an evolving complex language. Should we represent them as *tectonic plates* of the primary grounds that converge in building human identity, the process could be viewed as the interaction of a *cultural plate* mounted on another one that preceded it, i.e., the *neurobiological tectonic plate* of primal animal origin. Though the former tends to generate strata that are typically to condition full expression of the neurobiological/neurobehavioural tectonic plate, the latter persists with primal drivers of behaviour (feeding, survival or self-defence, social prevalence, territoriality, gregariousness, sexual drives) that condition behavioural responses under demanding conditions.

The friction between both dimensions or *evolutive ancestral plates*—the cultural or collective and the neurobiological ones—are part of the constant human search for individual and social balance and survival, of human manifest behaviours and internal tensions, dissatisfactions, and contradictions, as well as the continued 'struggle' of human communities throughout history to incorporate or replace diverse values and priorities of different ethnic cultures. The adaptive predominance—or mismatch—of one or the other is manifested in the various behaviours of the human species, whether adapted, dysfunctional, or pathological. Two basic phenomena condition the dynamics of this complex interaction: brain neuroplasticity and heritable epigenetic changes. Therefore, it is not surprising that the complexity of community interpersonal relationships is derived from diverse competition, dominance, or survival profiles.

The evolutionary process in the human species resulted in a brain composed of components that recognize different stages of evolution of the original 'cephalic ganglia'.

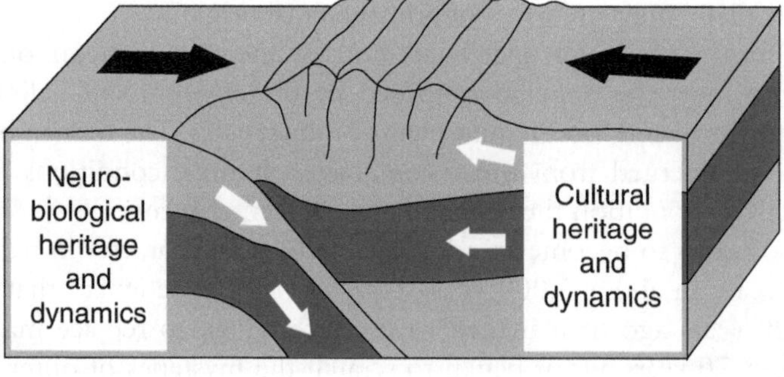


**FIGURE 4.1** Schematic illustration of the metaphor representing the neurobiological and cultural 'tectonic plates friction' as a source of adapted/non-adapted behaviours (from Colombo 2019, 2021a, 2021b).

It should be noted that such structures have been incorporated into new connectivity through evolutive stages, which condition brain processing. From a basic, physiological view, let us remember that the cognitive–socio–behavioural resultant cohabits with regulatory components of normothermia (intrinsic or behavioural thermal balance), normoglycaemia (feeding, appetites), reproductive sexuality, defence mechanisms (passive or active), and territoriality (physical space, public image). These represent the primal behavioural domains in human evolution and constitute the core of basal animal behaviour.

The friction between both 'plates' (or dimensions)—the cultural and the neurobiological—reflects our constant search for individual and social balance of our internal (states of mind) or manifest behaviours involving internal tensions, dissatisfactions, and contradictions. At the social level, these tensions are expressed as the continuous struggle of communities throughout history to incorporate or sustain differing human values and priorities.

> The complex debate resides on how social norms and rules could be avoided to come into conflict with the optimization of that creative capacity and initiative, say, with the genius of the human species. It is evident that the current conditions—so dissimilar between communities, so materialistic and consumerist, politically so hypocritical, and socially so inequitable—would require broad, progressive, cultural changes.
>
> *(Colombo, 2020)*

The analysis of these multiple variables results in the construction of complex relations between communities and nations. They are based on their technological capacity, educational standing, ecological incidence, vital statistics, and expansionist drive. These conditioners highlight the significant worldwide asymmetries in social construction, culture, knowledge, financial dealings, scientific-technological development, and the drive and ability to develop an aggressive or collaborative standing and to resist foreign menaces.

These factors make unreal the notion that we are living in a 'globalized' world but instead support the idea of an uneven, or a 'segmented' (or parcelled) world between the actors of human civilization, with transient periods of distended, convivial relationships as well as periods of struggle and of seeking the physical or cultural elimination of non-compatible interests. The profound inequalities and abuses involving peers and the environment hide either an undesirable by-product of the system's dynamics or a supposedly 'necessary' collateral damage involved in the concept of human evolution. This 'collateral damage' results in the marginalization of populations that do not consume and do not produce according to the needs of the 'market', whether in terms of the nature of the product or of sufficient quality or quantity for a world governed by the rules of supply and demand and of goods consumption. Another consequence is the transformation of some communities into passive consumers of food and other goods produced by corporations that control large sectors of the international market—based in more permissive, or less 'regulated', countries. In brief, the consequence is a society conformed by comparatively few leaders building a *manufacture of consent* for the world majority.

### Social and Cultural Darwinism or Social Construction and Multiculturalism

> In modern times, humans' unequal access to cognitive and technological development has provided grounds for furthering the social gap within and among communities and countries.
>
> ... social inequality merged under diverse socio-ecological conditions and would represent a universal component in the animal kingdom. Must our species passively proceed with such a backpack or, instead, attempt a cultural transformation?
>
> *(Colombo, 2022)*

The percentages of poverty across the world continue to demonstrate the existence of chronic poverty as an unsolved condition. The levels of exposure to hunger, diseases, and lack of stable employment, as well as the chronicity of such conditions, affect the developmental conditions of those communities. In turn, they affect survivability in a world increasingly segregated from those who have—or may have—access to the goods necessary for a competitive, dominant, productive style of life. These latter communities would include having access to comparatively high levels of education and a productive social insertion—in a creative sense, not only based on capital investment and commodities consumption. The development of knowledge and technology has placed pressure on communities' comparative ethical and emotional population structure, modifying or placing at stake traditional cultures and those that succumb to its expansion.

This accelerated 'population eugenics' or 'cultural eugenics' raises the need to define what humans aspire to as the generating base of social organization: whether it should be flagrant social Darwinism or social construction. The latter would include the concept of proactive solidarity and manifest equity, and its policies and actions are adjusted to those goals and cultural traditions. In this regard, the question can acquire a different dimension: are we considering the existence of two world populations—of two 'humanities', of two fundamental cultural 'phenotypes', one controlling and shaping social changes and the other with limited social mobility and condemned to the role of consumer or primary labour force?

The explicit tendencies extend to political and ideological domains, where suppression or silencing of the opponent becomes a primary goal. The interests for predominance in the distribution of political, economic, and military power have led to military invasions—or more subtle means of financial penetration and market control—with vested interests expressed under false pronouncements, the elimination of political leaders and activists, massacres of civilians, among other repressive actions. Thus, the continuous struggle for political/cultural survival or dominance has prolonged the *Darwinian arena* throughout human civilization. The objective has not been to foster *cohabitation* with what is different but rather to compete and fight it physically or functionally until its final extinction, obliteration, or subjection. This trend appears today potentiated by new technologies and alliances. The geopolitical projection of this drive for dominance implies a particular danger of uniformity of thought, accelerated by profound disparities in technological development. In this arena, political dominance exerted among countries expresses different strategies and results, either based on economic or military power. Through human behavioural history, this has consolidated a blatant, uneven, accelerated growth, with millions of human inhabitants displaced towards limited growth options or non-competitive, sub-human living conditions. Individual freedom is conditioned by agents of political or economic or religious power, who attempt to regulate the potential menace involved in public freedom of thought and speech, their growth, and their active social involvement. On evolutionary and comparative terms, animal dominant drives are expressed by controllable territorial limits that humans have expanded to include Earth and its space.

The multiple variables that go into constructing relationships between communities/nations, based on their technological capacity, educational coverage, ecological incidence, vital statistics, etc., do not suggest a 'globalized' planet. They highlight significant and worrying asymmetries in social, cultural, financial, military, and scientific-technological development, making unreal the notion that we live in a 'globalized' world. Instead, they support the idea of a 'segmentation' among the actors of human civilization.

The development of knowledge and technology has placed pressure on the structure of communities, modifying their traditional cultures and way of life, some of which are on the edge of extinction. Large sectors run behind each technological advance when humanity should be controlling such advances and subjecting them to criteria of distributive strategies, reason, and open opportunities. The profound inequities and abuses towards fellow human beings and the environment generate an undesirable by-product in the system's dynamics or collateral damage whose dimension has escaped control. This hypothetical collateral damage results in the socioeconomic marginalization of populations that do not consume and do not produce according to the needs of the 'market'.

Explicit technological trends invade the political and ideological arena: the fight for the extermination of the opponent. The interests for predominance in the distribution of political, economic, and military power have led to the continuous struggle for political/cultural survival and prolonged the Darwinian characteristics present throughout man's civilization: the objective has not been to coexist with what is different but dominance enhanced by new technologies and alliances.

Chronically impoverished communities would lack the necessary dynamics to escape their current situation and join the train of civilization that defines the new 'humanity'. This extremely Darwinian position—quasi-genocidal—deserves a separate paragraph since the effects of the policies and the continuous development of science and technology reinforce this trend toward the segmentation of humanity. A segmentation that, by its dynamics, leads to the cultural replacement—if not population replacement, a slow and undeclared physical/cultural genocide—of many communities whose natural imprint, often the product of a history of conquerors and oppressed, does not coincide with the 'democratic' and 'equal rights' declarations of the leaders of the industrialized world.

The vigour of culture is better characterized by its continuity over time and its respect for cultural diversity rather than its aggressiveness and territorial expansion. The history of civilization has shown the rise and fall of empires, reigns, and social systems. Nowadays, the expansionist vigour of economic policies and ambitious entrepreneurs, supported by technological advances, negatively affect the quality of life of communities due to the erasing of cultural traditions, labour abuse or slavery, poverty, social marginalization, reduction of the sustainable ecosystem, contamination of land and waterways, and desertification.

> Food packaging and utensils commonly contain up to 68 'forever chemicals' that carry possible health risks, with regulators potentially being unaware of the presence of many of them. Perfluoroalkyl and polyfluoroalkyl substances (PFAS) are a class of synthetic chemicals that are used to produce goods such as non-stick cookware and waterproof clothing. The bonds between the carbon and fluorine atoms in PFAS are so strong that it can take hundreds to thousands of years for them to break down. Many of these chemicals have been linked to harmful health outcomes, including cancer and reproductive and immune problems.
>
> *(New Scientist Daily, 2024*[1]*)*

Under these conditions, it is anticipated that the potential of such communities for their development in adapting to other forms of relationship and obtaining natural resources would be progressively compromised, as would their survival or aptitude for individual development even within their cultural context.

There is an unavoidable fact of reality: at one end of survival condition lie those most affected by chronic hunger or poverty, by inadequate upbringing, and by social marginality. Its recovery will no doubt depend on special programmes and public policies. In these cases, given different degrees of physical and mental conditions, in some, the level of individual and social deterioration is bordering on the irreducible. Therefore, given the profound mental and physical decay into which they have fallen, they would not be proficient in fighting for such rights. According to that vision, such communities would lack the necessary dynamics to escape from their current situation and join the ***bandwagon of civilization***. In other words, they would have lost their ties with 'modern humanity'.

Post-industrial societies are becoming increasingly dependent on material and technological culture to the point of 'embraining' [sic] them, i.e., incorporating both into basic cognitive constructions and conceptually becoming ***technological hybrids*** or ***technologically dependent***. This process takes place without the full awareness and clear understanding of the standard population, which is culturally bound to follow the technological stream.

The possible conceptual error of this position would be to accept as final what is either an expression of a cultural phenotype that conflicts with the predominance of technocratic or consumer trends or a consequence of inadequate socioeconomic policies that break 'the back of identity' or 'cultural dignity' of those communities. On historical neurocognitive grounds, this has been interpreted as behaviour responding to an intrinsic mental 'inferiority' implied in Broca´s mental scale (see discussion in Gould, 1996). Were these not the prejudicial premises that were intended to support slave-owning ideologies or used to justify social stratification until well into the twentieth century and are still present today?

This position raises another dimension of the problem, which is shared with conquest and cultural invasion. How to generate conditions for the coexistence of communities with disparate profiles of community structure, value priorities, and internal dynamics?

The strength of a culture is not characterized by its aggressiveness and territorial expansion, but by its continuity over time and its respect for cultural diversity. We should have learned from the history of civilization—the rise and fall of empires, kingdoms, and social systems. The difference is that today the expansionist strength of economic policies and ambitious entrepreneurs is supported by technological advances while improving the standard of living of a dominant minority and reducing the quality of life of increasingly numerous communities, 'retarded' by different mechanisms: cultural traditions of labour abuse or slavery, poverty, and social marginalization, reduction of the sustainable ecosystem and impact on cultures depending on primary resources, contamination of the land and waterways, and desertification impact on traditional cultures. Under these conditions, it is foreseeable that the potential of such communities for their future collective development will be progressively further compromised, as well as their cultural survival or fitness. It should also be considered that, in several of these communities, the most valued behaviour is usually emotional or affective, attached to personal values, and not the ownership of material goods, which makes them potentially more vulnerable to financial voracity.

In this reality, the industrialized world imposes conditions on the 'emerging' or 'underdeveloped' world. Communications, the war arsenal, the demands of an unbridled economy, and the inability to be self-sustaining without resorting to 'adjustment' policies and over-exploitation of natural and human resources, military and technological dominance do not build a promising future for communities attached to other cultural traditions, nor for natural life on the planet. Galeano (2010) strongly affirmed this when arguing that underdevelopment is not a stage of development *but its consequence*. The communities with lower economic well-being indices express biological compensation indices in both vital extremes (World Health Organization, 2013). Under this perspective of profound inequalities and cognitive growth rate (with long-term impact on cognitive development), certain statements, such as those from Goldman Sachs[2] on AI (artificial intelligence) and its effects on human development, would seem projected to certain societies that could meet the required standards and undoubtedly distant from the reality of the almost one-third of world population that lives *at ground* or *below ground level* as mentioned in Colombo, 2020, 2022, 2024, listed below:

> ... roughly two-thirds of current jobs are exposed to some degree of AI automation, and that generative AI could substitute up to one-fourth of current work. Extrapolating our estimates globally suggests that generative AI could expose the equivalent of 300mn full-time jobs to automation.
>
> *(Colombo, 2020)*

> The good news is that worker displacement from automation has historically been offset by creation of new jobs, and the emergence of new occupations following technological innovations accounts for most of the long-run employment growth.
>
> *(Colombo, 2022)*

Globally, 18% of work could be automated by AI.

*(Colombo, 2024)*

So, once again, highly advanced technological societies progressively detach themselves from the non-technically driven or technologically updated 'rest of the world', reinforcing the concept of a fractured world society with radically different rates of cultural drives and social developmental profiles.

## Notes

1 *New Scientist Daily*, 2024. https://www.newscientist.com/article/2422925forever-chemicals-have-infiltrated-food-packaging-on-awidescale/?utm_source=nsday&utm_medium=email&utm_campaign=nsd ay_200324&utm_term=Newsletter%20NSDAY_Daily.
2 'The potentially large effects of artificial intelligence on economic growth' (J Briggs/D Kodnani). Goldman Sachs, 2023. https://www.gspublishing.com/content/research/en/reports/2023/03/27/d64e052b-0f6e-45d7-967b-d7be35fabd16.htm.

## References

Colombo JA. 2019. *Our animal condition and social construction.* Nova Science Publishers. ISBN: 978-1-53615-357-6. eBook ISBN: 978-1-53615-3583.
Colombo JA. 2020. *Creativity, a profile for human species.* Cambridge Scholars Publishing.
Colombo JA. 2021a. Ancestral drives and the complex social behavior of human species. *Int J Primatol Res.* https://www.researchgate.net/publication/349768145.
Colombo JA. 2021b. *The homo within the sapiens.* Nova Science Publishers. ISBN 978-1-53618-373-3. https://tinyurl.com/3wbdchlc.
Colombo JA. 2022. *Dominance behavior: An evolutive and comparative perspective.* Springer Nature.
Colombo JA. 2024. *Human biome and human behaviour.* Routledge.
Galeano E. 2010. *Las venas abiertas de America Latina* (*The Open Veins of Latin America*). Siglo XXI.
Gould SJ. 1996. *The mismeasure of man.* WW Norton.
World Health Organization. 2013. *World Health Report.* WHO.

# 5

# BRAIN EVOLUTION AND ENVIRONMENTAL INTERACTIONS RESET INDIVIDUAL PROFILES AND REQUIREMENTS

Peters et al. (2004) stated that brain mechanisms prioritize regulating their own adenosine triphosphate (ATP) availability. In this regard, according to Krieger (1921; quoted by Hitze et al., 2010), the brain behaves in a 'privileged' manner, as evidenced by the human brain mass preservation during famine or inanition, while all the other organs of the body such as heart, liver, kidneys, and pancreas lose about 40 per cent of their mass. As analysed by Foley and Lee (1991) and Leonard and Robertson (1992), relative brain size has been linked to energy requirements. In this regard, anthropoid primates would spend about 8 per cent of their brainpower on their resting metabolism, a significantly larger proportion than other mammals (3–4 per cent) but still considerably less than the 20–25 per cent in adult humans. Shultz et al. (2012) proposed a correlation between brain size increase and cognitive development in *Homo* species based on demographic change and language development, to which probably material culture should be added as an additional interactive factor. Brain size increase in *Homo* has been a gradual process in Africa, which was completed following the migration of hominins into Eurasia associated with changes in brain size. In this regard, Aiello and Wheeler (1995) proposed the 'expensive tissue hypothesis' by which brain size changes would have occurred at the expense of the digestive system. However, this hypothesis has been challenged by Navarrete et al. (2011), who found that, controlling for fat-free body mass, brain size was not negatively correlated with the mass of the digestive tract or any other expensive organ, thus refuting the expensive tissue hypothesis. These authors proposed that human encephalization was possible by a combination of stabilization of energy inputs and redirection of energy from other sources. According to these authors:

> We propose that during human evolution improved diet quality, allomaternal subsidies, cognitive buffering, reduced locomotion costs and reduced allocation to production all operated simultaneously, thus enabling the extraordinary brain enlargement in our lineage.

DOI: 10.4324/9781003571544-6

Kuzawa et al. (2014), contributed to this discussion based on brain metabolic demands during childhood. Considering that brain glucose demand relates inversely to body growth from infancy to puberty, these authors concluded that the unusually high costs of human brain development require a compensatory slowing of childhood body growth.

As stated by Hitze et al. (2010), the energy-deprived brain can switch to an energy-sparing mode, which shuts down brain function to maintain cerebral energy homeostasis, which serves neuroprotection. In such a case, ATP-sensitive potassium channels in neurons in the brain stem and hemispheres are opened, and neurons are hyperpolarized following the intraneuronal decline of ATP concentrations, followed by a decrease in neuronal firing rate.

These processes are critical during developmental stages, such that early life experiences are vital for most animals' survival and healthy development, including humans. In this regard, during the extended human immature postnatal stage, disruption of normal parent–child interactions can have long-lasting effects on the quality of life of the offspring, as reported by Lai and Huang (2011). Additionally, such disruption would affect neuronal growth and maturation of the reward circuits that modify behaviour under reward-predicting conditions (Ryakiotakis et al., 2023). As a comparative example, rats that underwent a denial of expected maternal reward (expressed as maternal neglect) during early postnatal life developed in adulthood altered activation of the reward circuitry, anhedonia, aggressive play-fight behaviours, and aberrant prefrontal cortex structure and neurochemistry. According to Lai and Huang (2011), early life stress modifies the hypothalamic-pituitary-adrenal axis development, brain function, and neurobehaviour. Experiments conducted in rats applying post-weaning social isolation resulted in impaired social recognition and neuronal changes in involved neural circuitries (Guimarães et al., 2023). In essence, as stated by Salwiczek et al. (2012),

> … animals' cognitive abilities are linked to their evolutionary history, and hence their ecology.

On a neurobiological basis, as explored in mice by Yamamuro et al. (2020), sociability would involve a circuit composed of prefrontal projecting pathways to the posterior paraventricular thalamus. However, according to Xu et al. (2021), multiple microcircuits and input-output circuits would converge on the medial prefrontal cortex, hippocampus, and amygdala/accumbens nucleus to regulate complex social behaviours. Basic behavioural patterns are acquired early in vertebrates, during which social interactions provide essential wiring instructions affecting brain development (Atzil et al., 2018; Ilyka et al., 2021).

To place the issue in a social context, brain neurogenesis and neural circuit ensemble formation undergo developmental stages that depend on the species' socioecological (habitat) conditions. During this species-specific timing, the brain traverses several stages of susceptibility to either social or physical environmental factors. Thus, neural brain and mental activity constantly interact with environmental, emotional, and introspective domains. In this multivariate condition, sociality plays a significant role in their development and, to a considerable extent, in individual developmental cognitive and emotional profiles later expressed in adult life. In this regard, historical adaptation to the natural ecological niche incorporates a concept of relativity in terms of conditioning the development of cognitive resources. Hence, material, social, and emotional conditions during early development affect individual degrees of cognitive freedom and brain/mental profiles during adult life. Thus, relative social standing, enriched environment or poverty conditions, and social hierarchies are variables that could be considered universals based on their impact on brain development and cognitive performance in mammalian species, as summarized in Colombo (2019):

> Several experimental studies followed those initially performed by Harlow (1959) and the groups of Greenough and Rosenzweig (Krech et al., 1960; Volkmar and Greenough, 1972; Turner and Greenough, 1985). They confirm that the physical and social conditions of the environment during the breeding period and later affect brain development. This takes place at the microstructural and neurochemical level; a condition that influences cognitive abilities and emotional behavior in adult life.

The ecological domain should include the interaction of past cognitive behavioural profile responses and a proximate sociocultural environment that conditions brain neural plasticity. Interactions (dominant or prevalent vs. submissive) among individuals of animal social groups represent an intrinsic component of such 'environmental' or ecological context. Hence, a general overview would suggest that the evolution of *sapiens* in physical and behavioural domains resulted from a complex interaction of introgressed ancient behavioural profiles. This involved genetic and multiple environmental factors, all projected over time on a sort of evolutive neurobiological inertial conditioning. Thus, contemporary human neurocognitive profiles are not germane to the interactive history of the *sapiens* species. Behavioural success and failure provide a source of either reinforcement or decay of behavioural profiles. In this regard, contemporary behavioural components would recognize ancestral drives that caused friction with new environmental factors and culturally introspected demands. Hence, the notion that modern humans mature as if they were an initial *tabula rasa* would contradict the natural history of the human species. This poses an intriguing question, i.e., to what extent does the past evolutive history of *Homo* evolution remain ingrained in modern human neurocognitive ensembles and conditions underlying behavioural drives?

## An Excerpt on Tool Development in *Homo* Evolution

According to Shumaker et al. (2011), *tool use* could be defined as:

> The external employment of an unattached or manipulable attached environmental object to alter more efficiently the form, position, or condition of another object, another organism, or the user itself, when the user holds and directly manipulates the tool during or prior to use and is responsible for the proper and effective orientation of the tool.

The spontaneous occurrence of 'tool use' in the natural habitat of several taxa has been documented. In *Homo*, as quoted in Fragaszy and Mangalam (2018):

> Tools are ubiquitous in human cultures and have been central to human lifestyles since the origin of the genus Homo (Gibson and Ingold, 1994).

According to Semaw et al. (1997), 1.8-million-year-old (myr) stone-made artefacts of Oldowan assemblies were found at Olduvai Gorge (Tanzania), which adds to earlier findings at Omo (Ethiopia) and Turkana (Kenya) dated to 2.3 myr.

> The slow technological evolution of the lithic period, in terms of arrowheads and axes construction—from the Chatelperronian and Oldowan to Mousterian, passing by the Acheulian periods (Longa, 2009; O'Neil, 2012; Ruebens et al., 2015)—in a process that took approximately 2 million years, appears as the initial stages previous to the acceleration which occurred during the last ten thousand years, corresponding to the Upper Paleolithic period.
>
> *(Colombo, 2020)*

The previous comment excerpt should be contextualized within an evolutive context. As previously commented in Colombo (2020), the *Oldowan* and *Acheulian* tool traditions appeared during the Lower Palaeolithic stage, spanning between 3 million years and approximately 100 thousand years ago. During this period, at least three *Homo* species, *Homo habilis*, *Homo erectus*, and *Homo heidelbergensis*, would have roamed African and/or European landscapes. Additionally, the last two *Homo* species would have started Euro-Asiatic migration in such order of emergence and partially overlapping their presence in time. The later *Mousterian* lithic style would not have appeared until 100 thousand years ago when Neanderthals and modern *sapiens* species emerged.

Stout et al. (2008) explored neural correlates of early Stone Age making and considered that far from a breakthrough, in evolutionary terms, stone toolmaking would have been a fragile behaviour of marginal value. According to these authors:

> This is expected for a technology located near the limits of contemporary hominin capacities, with high learning costs, limited flexibility, and high rates of failure offsetting its benefits. Natural selection acting on hominin brains and bodies during this lengthy period of experimentation may eventually have eased these costs, leading to a dramatic proliferation of 'Oldowan' tool-sites after about 2.0 million years ago (mya). This is closely coincident with the first appearance of larger-brained and bodied *Homo erectus* by ~1.9 mya and was rapidly followed by the invention of more sophisticated Acheulean 'handaxe' technology by 1.76 mya, marking an important inflection point in the biocultural feedback processes that eventually produced the modern human technological niche.

Progressive complexity in toolmaking is not germane to a cultural complexity linked to brain/mental capabilities. In this regard, two opposing views contrast the timing for its progressive development. One supports the concept of a cultural '*big bang*', an acceleration that would have occurred 40–50 thousand years ago, perhaps associated with possible punctate genetic mutations, changes in feeding nutrients, and sedentary behaviours. An alternative view supports a cultural shift in pace, emerging as a slow, progressive process involving changes in cognitive development associated with a progressive increase in social, communicational, and quality feeding habits that modify brain chemical balance.

In its evolution since its emergence, *Homo sapiens* underwent several stages of survival and dependence: scavenger, prey, collector, predator, farmer, artisan or manufacturer, industrialist, employee, conqueror, imperialist, capitalist. In every one of these stages, since it began as a predator, *Homo sapiens* has been an explorer, innovator, and toolmaker of progressive complexity. This had consequences on the development of the mind-tool/instrument relationship, a combination that, together with the development of language and sociability, defined the basic profile of the species.

Many of these developments occurred under the influence of creative minorities and constructive, consuming majorities. The development of progressively more complex instruments and the notion of their future impact has generally remained alien to most people. This unequal awareness about the consequences of each development on the individual and collective future progressively expanded the cognitive gap between creatives, developers, instrument users and entrepreneurs, and most of the world's general population. In this process of creative and inertial expansion, *Homo sapiens* became an uncritical social modifier, a predator of species, and exploiter of mineral and ecological resources. It could almost be said that, in its development, *Homo sapiens* abandoned its integrated survival needs with the environment for its drive for profit and exploitation of resources; in short, a behaviour that would question its status as a communitarian *sapiens*.

Our successive distant ancestors evolved conditioned by a reduced series of resources. They included social adherence and the progressive development of various material means (tools) following their evolution and demands posed by interaction with members of the same species and with nature. It evolved from manufacturing arrowheads of different artisanal characteristics to complex research, development, communication, and dominance instruments. Regarding this proposed sequence, the role of early stone toolmaker has been attributed to *Homo habilis*, following the findings by Louis and Mary Leakey at Olduvai Gorge in Tanzania (review in Marzke, 2013). However, basic (raw material) tool use has been described in non-human primates and non-primate species (as in tailorbirds manufacturing 'pouches' to make their nests, and New Caledonian crows foraging with tools).

This early adherence by early *Homo* to the development of tools of different complexity and purposes distinguishes the vast range of valuable purposes related to varying stages of cultural development.

## References

Aiello LC, Wheeler P. 1995. The expensive-tissue hypothesis—The brain and the digestive system in human and primate evolution. *Curr Anthropol* 36: 199–221.

Atzil S, Gao W, Fradkin I, Barrett LF. 2018. Growing a social brain. *Nat Hum Behav* 2: 624–636.

Colombo JA. 2019. *Our animal condition and social construction.* Nova Science Publishers. ISBN: 978-1-53615-357-6. eBook ISBN: 978-1-53615-3583.

Colombo JA. 2020. *Creativity, a profile for human species.* Cambridge Scholars Publishing.

Foley RA, Lee PC. 1991. Ecology and energetics of encephalization in hominid evolution. *Philos Trans Roy Soc B* 334: 223–231. https://doi.org/10.1098/rstb.1991.0111.

Fragaszy DM, Mangalam M. 2018. Tooling. *Advances in the Study of Behavior.* Vol. 50. Elsevier, pp. 177–241. ISSN 0065-3454.

Guimarães DM, Valério-Gomes B, Vianna-Barbosa RJ, Oliveira W, et al. 2023. Social isolation leads to mild social recognition impairment and losses in brain cellularity. *Brain Structure and Function* 228: 2051–2066.

Hitze B, Hubold C, van Dyken R, Schlichting K, et al. 2010. How the selfish brain organizes its supply and demand. *Front Neuroenerg* 2: 7. doi: 10.3389/fnene.2010.00007.

Ilyka D, Johnson MH, Lloyd-Fox S. 2021. Infant social interactions and brain development: A systematic review. *Neurosci Biobehav Rev* 130: 448–469. doi: 10.1016/j.neubiorev.2021.09.001.

Krieger, M. 1921. On the atrophy on human organs in inanition. *Z Angew Anat Konstitutionsl* 7: 87–134.

Kuzawa CW, Chugani HAT, Gossman LI, Lipich L, et al. 2014. Metabolic costs and evolutionary implications of human brain development. *PNAS* 111: 13010–13015. https://doi.org/10.1073/pnas.1323099111.

Lai MC, Huang LT. 2011. Effects of early life stress on neuroendocrine and neurobehavior: Mechanisms and implications. *Pediat Neonatol* 52: 122–129. doi: 10.1016/j.pedneo.2011.03.008.

Leonard WR, Robertson ML. 1992. Nutritional requirements and human evolution: A bioenergetics model. *Am J Hum Biol* 4: 179–195. doi: 10.1002/ajhb.1310040204.

Marzke MW. 2013. Tool making, hand morphology and fossil hominins. *Philos Trans R Soc B* 368. https://doi.org/10.1098/rstb.2012.0414

Navarrete A, van Schaik CP, Isler K. 2011. Energetics and the evolution of human brain size. *Nature* 480: 91–93.

Peters A, Schweiger U, Pellerin L, Hubold C, et al. 2004. The selfish brain: Competition for energy resources. *Neurosci Biobehav Rev* 28: 143–180.

Ryakiotakis E, Fousfouka D, Stamatakis A. 2023. Maternal neglect alters reward-anticipatory behavior, social status stability, and reward circuit activation in adult male rats. *Front Neurosci* 17:1201345. doi: 10.3389/fnins.2023.1201345.

Salwiczek LH, Pretot L, Demarta L, Proctor D. et al. 2012. Adult cleaner wrasse outperform capuchin monkeys, chimpanzees, and orang-utans in a complex foraging task derived from cleaner–client reef fish cooperation. *PLoS One* 7: 1–9, e49068. doi: 10.1371/journal.pone.0049068.

Semaw S, Renne P, Harris JW, Feibel CS, et al. 1997. 2.5 million-year-old stone tools from Gona, Ethiopia. *Nature* 385: 333–336. doi: 10.1038/385333a0.

Shultz S, Nelson E, Dunbar RIM. 2012. Hominin cognitive evolution: Identifying patterns and processes in the fossil and archaeological record. *Philos Trans R Soc B.* https://doi.org/10.1098/rstb.2012.0115.

Shumaker RW, Walkup KR, Beck BB. 2011. *Animal tool behavior: The use and manufacture of tools by animals.* Johns Hopkins University Press.

Stout D, Toth N, Schick K, Chaminade T. 2008. Neural correlates of early Stone Age tool-making: Technology, language, and cognition in human evolution. *Philos Trans R Soc B* 363: 1939–1949.

Xu S, Jiang M, Liu X, Sun Y, et al. 2021. Neural circuits for social interactions: From microcircuits to input-output circuits. *Front Neural Circuit* 15. https://doi.org/10.3389/fncir.2021.768294.

Yamamuro K, Bicks LK, Leventhal MB, Kato D, et al. 2020. A prefrontal–paraventricular thalamus circuit requires juvenile social experience to regulate adult sociability in mice. *Nat Neurosci* 23: 1240–1252.

# 6

# EVOLUTION AND SOCIAL INEQUALITY

> The problem of inequality is relational, not economic.
>
> (Payne et al., 2017)

The quotation above poses a perspective on the source of inequality. Here, the authors have attempted to include an evolutive, comparative perspective of the inequality component in social construction. It limits social group adaptation and interactions based on individual and collective behavioural and physiological profiles.

Within the dynamics of gregarious social structures, the notion of a 'different' individual relates to the probability and success of an individual ascending or descending within the hierarchy of those social structures and, jointly, the probability of establishing some form of dependence or dominance in interpersonal relationships. The concept that 'inequality is the product of the fundamental laws of capitalism' (Piketty, 2014) would appear to ignore comparative studies on the social structure and dominance among non-human gregarious animal communities, where capitalism could hardly be invoked, as well as early stages in human feeding behaviours (e.g., hunter-gatherers and collectors). Species and ecological niche characteristics are conditioning factors that usually define hierarchical success or subordination and social marginals. In other words, in *non-egalitarian,* hierarchically based animal communities, social organization profiles, relative physical strength and endurance, and, in humans, diverse practical, social, and mental abilities are sources of unequal development and individual growth opportunities that condition their social status and relative dominant/submissive roles.

Among humans, animal ancestral drives are manifested or masked under different versions. However, institutional rules—designed by dominant structures, 'castes', or pseudo-democratic political consensus—confer a sense of rationality and order in social organization. At any rate, individual characteristics should generate the interactive, constructive, social 'mosaic' and not become a source of social deformations that cancel equal development opportunities. However utopic it could sound, it is '*demandable*' that

DOI: 10.4324/9781003571544-7

humans manage to generate trends towards 'utopias' that overcome cultural conditioning, which leads to authoritarian or pseudo-democratic social constructions (the latter based on conditioned freedom or access to education and information). In most societies of the modern world, material and behavioural drives tend to be moulded by consumerist practices and market conditionings.

Each historical form of sociopolitical organization likely impacted their citizens' neurobiological/neurocognitive, genetic, and phenotypic profiles and expressions. Beyond the formal differences in their social structure and relations between individuals and institutions, the engine that feeds them continues to be fuelled by several behavioural profiles. These involve dominance, territoriality, predatory cultures, and reassurance of a genetic and cultural predominance divided into castes or 'dominant groups'—financial, religious, corporate—, exerting dominant positions.

Inequality in social and feeding rights or opportunities is a current component in human and non-human animal communities, which could be considered a 'natural inequality' among gregarious animal communities with a hierarchical structure. Through eons and historical events, these ancient animal drives have been passed on to human social interactions, camouflaged under vicious rules that secretly or openly modulate the conditions of social interactions.

Social inequity varies depending on cultural and socioeconomic status conditioners. This raises an issue of 'fundamental animal conditionality' as 'naturally inserted' in human social structures that underwent historical and cultural adaptations. In this regard, Garfield et al. (2023) stated:

> Some norms, such as those promoting prosocial behaviour and shunning antisocial behaviour within groups, seem to be cross-culturally universal (Bowles & Gintis, 2002; Chudek & Henrich, 2011; Sober& Wilson, 1998). Violations even of universal norms, however, are pervasive. Unsurprisingly, punishment of norm violations is also a human universal and is implicated as a key mechanism promoting our species' distinct cooperative propensities (Boyd, Gintis, Bowles, & Richerson, 2003; Henrich et al., 2006). Substantial work has identified the mechanisms by which punishment can stabilize cooperation and incentivize prosocial behaviour.

The concept of social punishment draws on the concept of fear as a social controller, as stated by Boehm (2017, 2018): 'Humans' conscience functions are at least in part based on simple fear of punishment, and in both cases, fear leads to self-control.'

The above concepts bring up a crucial dilemma in social construction. Since escaping the fear of social repression would free up questioning the prevailing norms in social behaviour, it would be at the threshold for promoting either prosociality progress behaviours or plain delinquent behaviour. Within this realm, the quest to enhance conscientious behaviour responsibility must be considered instead of fear-based behaviour as a social controller.

Using an anthropological comparative evolutionary approach, what impact does inequality have on individual profile development, sociobiological interactions, and global interaction imposed by the conditionality of the personal and collective futures? Among public political structures vested interests linked to economic, financial, religious, and ideological interests and dominant powers based on their material possessions drive their dominant behaviour on crucial issues, such as conditioning equal rights to education access, freedom of speech, and social improvement. Critical knowledge and degrees of freedom are fundamental to overcoming such conditioning, where access to early education becomes the key to any substantial progress.

Let us consider the predominant examples provided by natural history. Throughout human history, we have faced a pre-eminent drive similarly construed based on a series of factors. Basically they relate to the role of hierarchies, social inequality, destruction of the opponent or potential threat, cultural predation, violence associated with maladaptation, violence in the imposition of the rules, conditioning of the *degrees of freedom*, educational priorities, privileged minorities, alliances of interests between power factors, demonstration of power with dissuasive purposes, and beliefs turned into fundamentalisms. As stated earlier (Colombo, 2019):

> At a human level, this uncontrollable sliding of increasing 'power force' manifests itself and perhaps is fully expressed in the recent proposals to generate a 'space army force' and to consider future extra-terrestrial mining (Lovett, 2012). In the meantime, vast population segments are condemned to material and cognitive impoverishment, or marginalization from knowledge (Colombo, 2015).

Cognitive development and emotional interaction among humans have been enhanced by the emergence and evolution of sophisticated language, which has allowed the development of complex virtual constructs based on symbolism and the instruments of culture, which tend to enhance inner thoughts and complex communication. However, all of them are supported by processes anchored in ancient neural networks distributed within cortical and subcortical interactions, with essential elements preserved during the species' evolutionary development and dependent on environmental and nutritional conditions during developmental phases. These environmental vectors condition the probabilities of attaining different cognitive levels and competitive behavioural profiles.

The prolonged human developmental period is replicated with each newborn, thus providing continuous instances for early individual-cultural interactions. Hence, in addition to the neurobiological substrate anchoring ancient emotional drives and profiles that underlie *Homo sapiens*' cognitive development, this remains open to cultural and emotional evolutive processes. On evolutive and comparative views, in the human case, it is from a relatively highly conscious species with instruments and technology able to propose different paths or outcomes of the cognitive evolutive process. This compounds the human species' fundamental responsibility involving ethical and social demands relative to our collective present and future construction.

The worldwide cultural domain of the anatomically modern *Homo sapiens* depends on the spectrum of phenotypes and ethnic groups that conditions genetic inheritance, cultural environment and behavioural expression. It is a reminder that social and environmental ecology are sources of factors that affect neurocognitive and emotional organization and individual cognition and social behaviour. This interaction ought to be placed within a larger realm, as stated in Colombo (2013):

> [W]e must assume that we are biological entities and therefore belong to the Natural Kingdom, with creative potential and symbolic thought. The amazing thing about it is that such potential has emerged from the evolutionary dynamic—from the existing materials in nature, in which we must recognize our fossil ancestors.

The characterization noted above could be summarized in the schemata mentioned before, composed of *biological and cultural tectonic plates* (see Figure 4.1) on the understanding that the development of the human species emerged from a highly interactive, natural evolutionary framework. Implicit in this construction are our cognitive wealth, limitations, and ambivalences, products of the friction between what has been schematically characterized as *tectonic plates*. This process progressively evolved into a dominant species among biological macro-environments (Colombo, 2010, 2017).

Coupled with the prolonged socialization/domestication process, emotional bias engendered the construction of supernatural agents initially based on a profound sense of fear towards the powerful notion of dependency on an unknown, overpowering knowledge, and acceptance of its nature. This dependence engendered rituals somehow attempting to bridge such a gap, expressed by behaviours aimed at controlling the feeling of helplessness and anxiety caused by phenomena whose nature was not entirely understood or controllable. This 'second universe' of emotional nature generated a sense of social bonding, shared protection, and a source of familiar promises and prayers that eventually involved human sacrifices in fantastic constructions that attempted to reduce the formidable power of natural events. In a neurobiological context, neural circuits associated with the notion of the 'acting self'—self-conscience—(Crivelli and Balconi, 2017) overlap primarily with those involved in the 'social brain' and the attribution of intentionality to extrapersonal acting agents.

Considering the previous comments, perhaps the fundamental challenge that the human species must face, besides its survival in a context of positive, constructive interaction with the social and ecological environment, would be to recognize its fundamental animal primal condition and, from such an understanding, build the means to increase its *degrees of creative freedom* (Colombo, 2013) and cooperative social construction.

The process of civilization has involved self-induced human domestication, which includes psychophysical and cultural effects.

Some authors take these observations as indicators that civilization has led humans to conditions of self-domestication, with its physical and behavioural correlates. Although the data does not allow definitive conclusions about the anthropometric evolution of modern *Homo sapiens*, the statement would not be without interest in the social and scientific debate regarding the evolutionary impact of current living conditions—even with their varying levels of quality of life. Brüne (2007) took the theme of human domestication and projected it onto the field of eugenics. He considered, in accordance with current evolutionary biology concepts, that physical and mental characteristics represent biological adaptations and, consequently, there is a broad spectrum of phenotypes. In the case of specific pathologies and neuropsychiatric alterations, he assigned importance to imbalances between the individual and the environment, or produced by agents of the latter, and elaborated the so-called 'diseases of civilization' as a product of a phenotype altered by the conditions imposed by the sociocultural environment.

Therefore, before discussing eugenic policies—except for cases of verifiable genetic, hereditable pathology—it would be necessary to reassess the living conditions imposed by societies. According to Allen (1997), genetic arguments are often used to mask the actual causes—social and economic—of alterations in human behaviour. Although the data does not allow definitive conclusions about the anthropometric evolution of modern *Homo sapiens*, it could be agreed that the statement is not without interest in the social and scientific debate regarding the evolutionary impact of current living conditions— even with their varying levels of quality of life.

Ideas linked to eugenic proposals have an early origin, among which xenophobic movements that attributed virtues or disabilities to different gene pools were not absent. Glad (2006) and Brüne (2007) provided a historical approach to the problem, although with different emphases. Glad supported the principles and the need to implement eugenic policies based on promoting the fertility of the genetically fittest and discouraging those who are genetically disadvantaged, and supported segregation in these terms ('like breeds like'). This position fails to consider or distinguish two fundamental aspects. In the first place, the notion of fittest is not an absolute category but a relative one, with criteria or objectives to be specified. Second, pathologies due to genetic alterations should not be confused with those that are the product of maladaptation to conditions imposed by a given society or civilization or due to sociocultural characteristics during a historical period of a given community. In addition, as Brüne (2007) rightly stated, some pathologies do not invalidate the capacity for creative production (as would have been the case of several 'geniuses' in the arts and sciences), nor the capacity for social interaction. Brüne, on the contrary, took the concepts of social self-domestication and social Darwinism as examples of the deviation that scientific knowledge can suffer when it is transferred—by politicians or technocrats—to social policies, attributing to genetic conditions what it is actually the impact of environmental conditions. This deviation is defined as *a naturalistic fallacy*. Let us bear in mind that the phenotype of many populations has been affected by extreme conditions, to a greater or lesser degree, of persistent social marginalization arising from sociopolitical conceptions.

> Fertility is declining globally, with rates in more than half of all countries and territories in 2021 below replacement level. Trends since 2000 show considerable heterogeneity in the steepness of declines, and only a small number of countries experienced even a slight fertility rebound after their lowest observed rate, with none reaching replacement level ... These changes will have far-reaching economic and societal consequences due to ageing populations and declining workforces in higher-income countries, combined with an increasing share of live births among the already poorest regions of the world.
>
> *(Bhattacharjee et al., 2024)*

Therefore, in this debate, several disturbing circumstances require specific considerations. Among them, what would be the selection criteria—an even more complex decision as it would take place under social and economic distortion conditions—and who would make such a decision? A radical position would be that any decision of this nature is unacceptable and that communities must evolve according to their interactive characteristics, correcting the social and health risks due to poverty and the current access to education. In this respect, these are critical goals to be pursued. In any case, the notion of stationary phenotypes is unrealistic, even in the most developed societies, given the continuous modelling by environmental or migratory influences with variable combinations of genetic pools and behavioural profiles.

For this reason, the notion of stationary phenotypes represents an illusion if it is intended to isolate it from cultural interactions or the conditions imposed by the habitat. The inevitable interaction and integration will infiltrate elements from other cultures and generate new phenotypes. Therefore, to speak of an ideal, universal, 'frozen' phenotype again contradicts what represents a constant in our survival as a species: adaptation in the variety.

At the opposite extreme, one cannot help but be disturbed by the comments of John H. Campbell (1995; quoted in Glad, 2006), a biologist at the University of California, who affirmed that the opening of the genetic code (inheritance), like the printed circuit of radio, will allow humans to redesign themselves at will. This possibility has not escaped the most vigilant scientists and officials, though the general population may be less aware of the possible consequences. Among those possible consequences are the progressive installation of radicalized eugenics in the hands of social leaders who could apply criteria that are the product of a culture based on social inequality. Probably, in such conditions, this possibility of meddling with DNA would be implemented by an elite who would thus have the possibility of differentiating themselves evolutionarily, in a few generations, from the rest of the humans who will continue their evolution subject to the vagaries and conditions of the genetic and environmental pool. This induced self-evolution of *Homo sapiens* would engender another species, which has been called by some authors *Homo autocatalyticus* (Campbell, 1995; quoted in Glad 2006) (that is, self-induced), which would be in control of the conditions of the evolutionary process and therefore the direction of its evolution, eventually relegating the rest of humanity to the status of an archaic species. Indeed, it will not be in the hands of scientists to decide the application of this possibility—predictably viable from the methodological point of view—but in the hands of political establishments that, in the recent history of humankind, have already chosen other actions based on scientific-technological progress, with controversial consequences. Once its feasibility is confirmed, the issue becomes strictly moral for some and has advantages and geopolitical considerations for others. The scarcity of reservations humankind has shown regarding the implementation of power does not leave much room for doubt that this would become a real option. Any benefit for the species is a desirable, shareable objective if provided universally. However, it is also clear that this implies a series of considerations related to the future composition of the human population without distinction. A future that already today appears compromised or segmented in terms of access to scientific and technological development and training in terms of knowledge and competitiveness.

Supporters of eugenics maintain that it is based on four pillars: access to natural resources, sustainable biodiversity, a population that does not exceed what the planet can support indefinitely, and society's definition as formed by the population (Glad, 2006). Although the statements could be discussed, each one raises specific problems and contexts that are complex to solve if equity is incorporated as a significant variable. This complexity is also expressed in the proposals aimed at rationalizing—sharing at a planetary level—energy consumption rights.

## An Evolutive Perspective of Prosocial Behaviour

Should the spectrum of possible basic behavioural-survival profiles be placed under the standard frame of evolutive adaptations—considering survival as one primary, fundamental, and universal domain of life with its multiple evolutive profiles–, restrictive constructions of this essential component of life survival would probably be conceptually misplaced. Since early life forms provide an initial basis for the emergence of complex (multicellular) biological entities, associated nutritional and signal dependence profiles are required to organize a complex life form. Primal forms of *prosocial* behaviour included chemical signals that provided the basis for integrating multicellular ensembles. These shared available resources and physicochemical signalling on the state of the organized cellular conglomerate and unified—'prosocial'—behaviour against potentially menacing signal sources from the environment. So, if prosociality involves primary forms of mutualism and integrated cellular collective organizations, it would provide a basis for subsequent—more complex—life forms that evolved from these primal proto-organism entities. Through interaction with different and variable environmental conditions and subjected to various ecological conditions and survival and adaptive aptitudes, life forms evolved into diverse evolutive genera and species forms with variable adaptive and survival rates. The survival rate success was not foreign to the evolutive forms of prosocial behaviours. Though authors place a few restrictions on the concept, as posed by Kopp et al. (2023):

> Prosocial behaviours are not limited to humans but occur in a broad range of taxa and take place in contexts such as parental and alloparental care, provisioning of food, comforting, protection against predators, territory defence, and conflict resolution (Aureli and de Waal 2000; Clutton-Brock 2009; Jaeggi and Gurven 2013a; Burkart et al. 2017).
>
> Prosocial behaviours vary in effort and cost for actors.

Additionally, according to these authors:

> ... we define prosocial behaviour as 'voluntary, intentional behaviour that results in benefits for another' based on Eisenberg and Miller (Eisenberg and Miller 1987, p. 92). According to this definition, prosocial behaviour (or a prosocial action) is characterised by three features (see also Hawley 2014):
>
> I. The behaviour produces a benefit for one or more individuals other than the actor as a direct result of the actor's action, II. The behaviour is intentional,

> i.e. directed towards the respective outcome for the recipient(s), and not accidental, III. The behaviour is voluntary, i.e. the actor was not (noticeably) forced to act and would have been able to act otherwise ... [W]e are optimistic that it can also be expanded to other animal species.

If establishing a cellular consortium based on cellular affinity provides new properties to the conglomerate primal, then cellular conglomerates that provided the initial forms of prosociality would undoubtedly meet the first two conditions stated by Kopp et al. (2023). The concept of voluntary in the terms expressed by Kopp et al. (2023) implies inserting a cognitive process that does not seem mandatory in the natural kingdom for early cellular associations and organ/individual formation throughout evolution. Under the authors' scope, prosociality involves a cognitive process that evolved later in species evolution but was anticipated by primal cellular process signalling emerging into aggregation and mutualism in the complex process of evolution from cell aggregates to multisystemic organisms. Evolution generated a progressive inclusion of complex organic interactions and signals for systemic integration—always based on cellular processes—ranging from cellular signalling interactions to complex cognitive and behavioural dimensions, yet always based on cellular, systemic, and behavioural interactions. This last represents an exponential dimension of the primal evolutive origin, generating the emergence of new functional complex properties.

On prosociality, Wrangham (2021) considered that targeted conspiratorial killing

> ... is unknown in chimpanzees or any other vertebrate, and is significant because it permits coalitions to kill antisocial individuals cheaply. The hypothesis that major elements of groupishness are due to targeted conspiratorial killing helps explain why they are much more elaborated in humans than in other species.

In this regard, projecting onto the evolution of moral social control, Boehm (2018) stated earlier that:

> Collective intentions are inferable when they go on patrol, when they mob predators, when they go hunting, and when large coalitions gang-attack disliked members of the same community. This last behavior has particularly significant pre-adaptive implications for the evolution of moralistic social control, and it suggests that moral evolution has deep roots, going back to the Last Common Ancestor of humans, bonobos, and chimpanzees.

## Between Reality and Fiction

Knowledge to cultivate the land marked a significant social change: the settlement, the population growth, and the greater complexity of community structures. Socialization—a necessary process for the survival trial of a useless, big-headed hominin hatchling—was on its way to shaping a mind based on the complex interactions within and between groups. Under such conditions, human communities became progressively more numerous. Language is an instrument that allows creation of relational structures of various kinds (mystical, ideological, territorial, etc.), with the consequent segmentation of the population universe.

In communities organized unequally, individual skills and the formation of power groups deepened social stratification. With the increase in the size of communities, the notion of personal survival or 'equals' progressively modified the community reference of each member. The community structure began to fragment according to different parameters. In general, and in metaphorical terms, *Homo sapiens faber*, *Homo sapiens nundinator* (*Mercator*, merchant) and *Homo sapiens pauper* were distributed within increasingly skewed spaces of land and public income. Machiavellianism, deceit, manipulation or direct usurpation took centre stage in social relations and the exchange of labour effort or its product. Once the mechanisms of community hunting, gathering and distribution were abandoned, they were transformed into the barter of goods and then into a market for transactions and uneven accumulation, which soon ceased to have social objectives, and a just recognition of the effort made.

The Industrial Revolution introduced a gap between the producer's identity and what was produced, and communication technology helped propagate real or virtual financial and cultural empires. With such an avalanche and socioeconomic pressure, scattered population nuclei clung to local or regional traditions. The cumulative concept (of power, capital, means, knowledge) prevailed over the criteria of distribution and equity. At that time, the energy channelled towards primitive territoriality and the survival instinct emerged in complex behavioural structures, almost hiding its ancestral phylogenetic origin. It was not enough to have an extended cerebral cortex to silence our origins embedded in subcortical circuits and programmes. Such fundamental principles of animal behaviour as dominance, territoriality, competition, aggression, and survival continued to drive the sophisticated behaviour of this new primate that would finally find its expression in the form of national or supranational, oligopolistic, corporate political-financial organizations.

The enormous profits and wealth of the world's productive apparatus ceased to have a social role, becoming an instrument of socio-political-military dominance. The feuds ceased to have strictly local dimensions. *Homo sapiens nundinator* (merchant) began to win the battle. With severe and profound consequences, the notion of the market replaced the idea of community and humanity. The cultural segmentation, the division within the same community, generated dimensions of equals and strangers competing for territory and survival. It should not become so much of a surprise, then, that in current world communities, tribal traits are manifested in events that provide some symbol of identity to marginalized groups in society. The search for identity and shelter from the feeling of insecurity was satisfied with accessible physical means, or through distant magical, mystical, or esoteric sources. The sense of identity or of belonging to an affective, cultural, and ecological 'niche', and the traditions and legacies that sustained the ties of that community, began to fade, evaporating from personal representations and collective memory. That emptying was filled with silence, drugs, esoteric constructions, or degrading behaviours towards the environment.

The divorce between the exercise of power and the search for the common good has led to the alienation of political/financial structures from the general community. Today, the process of cultural 'homogenization'—functional for the consumerist objective sponsored by large enterprises—obtains strong allies in campaigns or in advertising and financial pressures, which maintain a window of absurd needs or promote 'food alchemies', where the essential is confused with the banal or superfluous, or progress in biotechnological knowledge, with campaigns destructive of biodiversity. It is the mercantilist universe of *Homo sapiens nundinator*. Excluded, the *Homo sapiens pauper* wanders between submission, criminal violence and revolutionary zeal. Divided, on the one hand, the concept of the planet prevails as a great 'bowling alley' or 'kiosk' and of its inhabitants as a malleable set of potential, virtual or actual consumers. On the other hand, solidarity utopias or social exclusion emerge.

If the final, absolute value of the principles of the prevailing communities ('fittest'?) is the survival of the species regardless of the values it supports or the conditions in which this occurs, then the concentration of power and the irrepressible scientific-technological progress—a powerful instrument of knowledge, turned into an instrument of domination—will continue its almost omnipotent development beyond any natural, social or ethical consideration. If it continues to be so, the fragmentation of the human community will undergo an irreversible change. By then, these communities may have been excluded from the orbit of natural and genuinely communitarian processes and have generated their own evolutive game rules. This process may have already started: a self-controlled and anthropocentric evolution with supranational, very long-term anthropocentric projects, which will probably take them to Alpha Centauri in the frame of ancient exploration drives. It is not easy to imagine the degree of individual freedom, how that human convention would work, and if humans would recognize themselves in them.

But also, perhaps, in some parcel of our terrestrial crust, some remnant human community of the old, relegated *Homo sapiens* will once again govern itself with the council of elders and be linked to the events of the natural kingdom—which will stubbornly 'try' to regenerate biodiversity—, while some artist would recreate aspects of their daily life in another Altamira Cave.

With the *impronta* of our primitive ancestors on our backs and our prodigious mental organ with the capacity for strategic, creative, and Machiavellian thinking on our shoulders, what can we do for the planet and its inhabitants? Given a new opportunity, will humans perhaps exercise their enormous potential in other directions?

We are faced with a severe dilemma. Either the human species continues current trends and represents another endpoint of this formidable natural story, or humans accept that their capacities should not nullify the role of sustaining the biodiversity of life on the planet and its phenotypic variety of species. In that case, human civilization could continue adding pages to such an imaginary book. Otherwise, it must come to the unfortunate conclusion that in this endless series of beautiful adaptations to the external environment represented by dynamic biodiversity, with its successes and failures, humans represent *the wrong species* of the natural kingdom (Colombo, 2010). Or, finally, that it carries the seeds of its own extinction.

Furthermore, one might wonder if we, the human species, have not contributed to undermining the concept that life is an open opportunity for another in which our actions are nothing more than the expression that we constitute an actual plague, capable of surviving—given enough time—in any circumstance and habitat, no matter how extreme, and prone to eliminate/cancel the remaining species, either directly or indirectly.

Today, actions that tend to break the natural balance are potentiated by human actions. Humans seem to constitute a plague that struggles to transform the universe so that it becomes accessible to its image and needs. It is highly desirable that the acceleration in technological development and its application by the—minority or elitist—leaders of the time, generally insensitive or self-satisfied, will not cancel the human collective future. If the genome does not modify the expression of the drive for intra- and interspecific dominance, humans could end up being replaced—by assimilation or expulsion—by another 'species' that expresses priorities and viable behaviours able to coexist and show respect for their congeners and the planet's biodiversity. In such a case, the emergence of this *Homo sapiens sapiens verum* will face a formidable opponent, armed with its 'parochial gene' and technological support. For this reason, it might be desirable to have a slow process of extinction of the human species and develop by assimilation into a species variant, a process for which there is no reason to rule out that it has already begun. It is about *our conquest by 'other' us.* Choi and Bowles (2007) analyzed the possible co-evolution of altruism and war behaviours and conclude that in primitive human communities, those that combined both altruism towards their companions and parochial hostility towards the stranger probably prevailed. Beyond the relative predominance of one imprint or another 'altruistic' or 'hostile', its final expression will be favoured or interfered with—i.e., conditioned—by the individual's cultural 'niche' from its early stages of development.

Let us suppose it is accepted that phenotypic and cultural varieties are inseparable components of the central motor of the adaptive capacity of *Homo sapiens.* In that case, the current conditions constitute a flagrant transgression of such capacity since it is generating a society where almost half of the world population faces social immobility and homogenization conditions due to deprivation of the elementary conditions for individual growth and its development or of the conditions that spark interest in such a developmental vector. This constitutes an immoral condition—if we echo a morality of solidarity—and intrinsically counter-evolutive—to the extent that it contradicts the essential need for infantile primary containment and reduces the potential gene pool. Perhaps the path that the human species has been following configures a kind of series of reiterative actions towards the extinction of *sapiens* as we know it today. What would evolve then?

Let us imagine for a moment the existence of rigid patterns, immutable in time, of a moral nature, of physical aptitude, of eating habits, of uniform social or political structures, and of cultural peculiarities. Perhaps then we could not imagine a worse planet for humans, a uniform and rigid planet, although also more predictable—in this case, at the cost of individual and collective creativeness—a significant value in the characterization of *Homo sapiens.*

The adaptability and plasticity of the species depend on several factors. Still, it thrives on variety and the possibility of expressing a range of physical and cultural phenotypes under ever-changing environmental conditions—quite different from isolated, primitive cultures. Such freedom broadens the possible spectrum of the trial-and-error game that characterizes the engine of natural evolution. The reduction in the degrees of cultural freedom would make humans and their communities more fragile by limiting their adaptive capacity and the eventual emergence of suitable and successful phenotypes in continuing to provide human civilization with original approaches and solutions to their problems and needs. The lack of plasticity could only generate more violence—either to try to survive or to dominate—directed towards different cultures, the environment, or their community, as has happened on many occasions during human history.

We should have learned enough from our war history from early times to today. With time and the development of human abilities, the warrior has replaced the hunter. Today, it is not the variety or the difference that places us at risk; it is the validity of proposals of unique, exclusive, hegemonic thought—political or religious. Biotechnological and scientific advances should promote productive, proactive policies towards a viable, multicultural, supportive and equitable humanity and planet—instead of transforming it into a barren field with remains of battles for temporary supremacy or a unilateral and unsustainable socioeconomic development.

We have considered a series of current and projected circumstances relative to the physical and environmental conditions that anticipate an unfavourable course for the continuity of biodiversity and the ecological conditions necessary for human survival. Its violation is observed with negligence by sectors of the ruling class at a global and national level. Their actions create an antinomy with the efforts of other sectors of society committed to providing theoretical and operational resources to change the direction chosen by the opponents or to try to ensure that sensitivity, aesthetics, and creative imagination continue to be part of our human profile. History indicates that leaders with financial, military, or political power aspire to have technology and knowledge within their reach, in order to continue enjoying impunity and privileges, and a dormant, passive, or underdeveloped population. The slow emergence of a *new human* must—perhaps—go through demanding survival vicissitudes like those our ancestors went through, thanks to whose inventiveness, tenacity, and communal sense humans have so far survived. The strategy to achieve that *new man* will be outlined and carved in the daily future, with its struggles, hopes, failures, and disappointments.

## References


Allen GE. 1997. The social and economic origins of genetic determinism: A case history of the American Eugenics Movement, 1900–1940 and its lessons for today. *Genetica* 99: 77–88. doi: 10.1007/BF02259511.

Bhattacharjee NV, Schumacher AE, Aali A, Abate YH, et al. 2024. Global fertility in 204 countries and territories, 1950–2021, with forecasts to 2100: A comprehensive demographic analysis for the Global Burden of Disease Study 2021. *Lancet* 403: 2057–2099. doi: https://dpoi.org/10-1016/S=140-6736(24)00550-6.

Boehm C. 2017. Ancestral precursors, social control, and social selection in the evolution of morals. In MN Muller, RW Wrangham, and D R Pilbeam (editors), *Chimpanzees and human evolution*. Belknap Press, pp. 746–790. https://doi.org/10.4159/9780674982642-020.

Boehm C. 2018. Collective intentionality: A basic and early component of moral evolution. *Philos Psychol* 31: 680–702. https://doi.org/10.1080/09515089.2018.1486607.

Brüne M. 2007. On human self-domestication, psychiatry, and eugenics. *PEHM* 2: 1–9.

Campbell JH. 1995. The moral imperative of our future evolution. In JH Campbell, R Wesson and P Williams (editors), *Evolution and human values.* Rodopi Press, pp. 79–114. www.home.comcast.net/~neoeugenics/camp.htm

Choi J-K, Bowles S. 2007. The coevolution of parochial altruism and war. *Science* 318: 636640.

Colombo JA. 2010. *Somos la especie equivocada?* (*Are humans the mistaken species?*). Eudeba. https://tinyurl.com/lsdozicg.

Colombo, JA. 2013. *Bajo libertad condicionada* (*Under conditioned freedom*). Imago Mundi.

Colombo JA. 2017. *Creatividad o domesticación* (*Creativity or domestication*). Buenos Aires Books.

Colombo JA. 2019. *Our animal condition and social construction.* Nova Science Publishers. ISBN: 978-1-53615-357-6. eBook ISBN: 978-1-53615-3583.

Crivelli D, Balconi M. 2017. The agent brain: A review of non-invasive brain stimulation studies on sensing agency. *Front Behav Neurosci* 11. http//doi.org10.3389/fnbeh.2017.00229.

Garfield ZH, Ringen EJ, Buckner W, Medupe D, et al. 2023. Norm violations and punishments across human societies. *Evol Hum Sci* 5: e11. doi: 10.1017/ehs.2023.7.

Glad J. 2006. *Future human evolution: Eugenics in the twenty-first century.* Hermitage Publishers.

Kopp KS, Kanngiesser P, Brugger RK, Daum MM, et al. 2023. The proximate regulation of prosocial behaviour: Towards a conceptual framework for comparative research. *Anim Cogn* 27: 5–15.

Payne BK, Brown-Ianuzzi JL, Hannay JW. 2017. Economic inequality increases risk taking. *PNAS* 114: 4643–4648.

Piketty T. 2014. Capital in the twenty-first century: A multidimensional approach to the history of capital and social classes. *Br J Sociol* 65(4).

Wrangham RW. 2021. Targeted conspiratorial killing, human self-domestication and the evolution of groupishness. *Evol Hum Sci* 3(e26): 1–21. doi:10.1017/ehs.2021.20.

# 7

# PRIMATE BEHAVIOURAL EVOLUTION

## Its Imprinting on *Sapiens* Behaviour

In the natural kingdom, all species are programmed to secure self-survival, which involves relative predominance based on a territory that offers nutrients and reproductive opportunities. Intrinsically, this behaviour—specific for each species and ecological niche—favours maximizing the probability of success and minimizing the likelihood of extinction or individual/species replacement. From these fundamental evolutive premises emerged the human species, exerted through eons of evolutive processing of basic behavioural mechanisms, though with species and phenotypic variations. This basic, evolutive construction has been largely hidden under human creations of cultural values and beliefs that mask the presence of those ancestral animal drives. Hence, human sociocultural systems attempt to reformulate, sublimate, and camouflage ancient, primary animal drives.

The neural circuits linked to emotional behaviour include components associated with violence, typically involved with self-defence, social prevalence, reproductive behaviour, territoriality, and access to food resources. This is exerted on an individual and a group collective basis and has an evolutionary resemblance with the group behaviour of several other animal species that regularly occur as killing coalitions. According to the imbalance of power hypothesis, these are attributed to the expression of survival and dominant drives. Among humans, such inter- and intragroup competence would take place on a similar basis, though masked behind cultural and belief paradigms. Hence, it can be triggered by survival or cultural reasons (religious, financial, ideological, or the expression of tribal parochialism associated with various possible slogans). The final expression of emotional behaviour (whether aggression or flight) responds through a common, final neurobiological pathway. This behaviour among *Homo sapiens* evolved into an arms race that progressively expanded and involved the fate of public collective and standing-by communities.

In evolutive terms, this behavioural pattern has been compared with those of close evolutionary primate relatives. The common primate evolutive predecessor to *Homo*—the

DOI: 10.4324/9781003571544-8

ancestral *Pan* of hominid primates—and to the chimpanzee (*Pan troglodytes*) and the bonobo (*Pan paniscus*) would have lived in social constructions based on hierarchical dominance. Yet, conflict management tends to differ for each non-human primate species. According to Furuichi (2011), while chimpanzees tend to conflict with male predominance, bonobos would display preventive behaviour with female predominance. Boehm (2012) proposed an interesting hypothesis according to which humans would keep both of these alternatives. That is, a potential behavioural bipolarity with uneven prevalence distribution among individual characters and social organizations. This hypothetical essential behavioural duality could be linked to the configuration of the human genome and the subsequent neurobiological scaffolding. It could be expected that those basic tendencies would predominate in different groups and individuals, depending on cultural and behavioural conditioning. Their final behavioural phenotype would depend on sociocultural and sociobiological conditions that mould prevailing behavioural genotypes.

According to De Waal (2007), humans would have a bipolar characteristic:

> What is most critical for survival: competition or cooperation? Are we more like chimpanzees or bonobos? … Such questions are a waste of time for the bipolar characters that we are.

Yet, as posed by dos Santos and West (2018):

> [C]ooperation and cognition can coevolve—cooperation initially evolves, favouring enhanced cognition, which favours enhanced cooperation and stabilizes cooperation against a drop in relatedness. These results suggest that enhanced cognition could have transformed the nature of cooperative dilemmas faced by early humans, thereby explaining the maintenance of cooperation between unrelated partners.

If the final expression of the supposedly bipolar human nature depends on cultural conditions, the occasional predominance or expression of one or another would depend on public policies that contribute to the construction of social relationships. Besides, if it is considered that behaviours are induced or reinforced by their reiteration, such public policy decisions acquire a collective pedagogic significance. That is, moulding a society where such bipolarity could be reduced and express itself either as predominantly cooperative or competitive.

According to observations made by Cole (2009), socioenvironmental processes regulate human gene expression. They affect hormonal and neurotransmitter activity, which interact with cell receptors. These, in turn, activate genetic transcription factors. They also condition the fate of those genetic expressions that best adapt to conditions imposed by cultural demands. This plasticity stresses the tight relationship between genetic inheritance and socioecological variables. In this regard, Jablonka and Lamb (2007) proposed four types of heritability (evolution in four dimensions): genetic, epigenetic, behavioural, and symbolic, each providing variations over which natural selection would act. This biosocial interaction is essential to understanding a series of processes at the individual and collective levels in the brain and behavioural organization field.

In the complex interactive arena of species evolution, the above has occurred under changing environmental conditions and proximities of competing species for feeding resources and territory dominance. Thus, anthropoid species developed under a naturally violent context that challenged individual and species survival and imprinted and reinforced defensive and dominant behaviours. Hence, it is not completely surprising that, eons later, violent behaviour remains a basic expression in individual and collective human social interactions. This has a neurobiological substrate based on neural circuits involving limbic structures (amygdala, hypothalamus) and prefrontal cerebral cortex (Blair, 2004; Gobrogge et al., 2007; Siever, 2008), as well as hormonal influences (Rose et al., 1972; Gordon et al., 1979), with which sociocultural scaffolding interacts in the final construction—plastic within biological conditioners—of individual behavioural profiles.

Our evolution as a species and the development of new operational instances—intraspecies tactics and strategies—were made possible due to the reconfiguration of distributed, conditional neural nets. These were immersed in our neocortex's notable cell density and connectivity evolutive increase and its subcortical (bottom-up and top-down) connections (see, for example, Dunbar, 1993; Striedter, 2006). Also, *Homo* evolution would have taken a ride on the reformulation of programs motorized by the demand imposed by the high degree of socialization of the human species. That is, an extended ecological demand to incorporate the environment's physical, emotional, and cultural events. Adaptation to such socialization and brain demand, overlapping the high energetic requirements of the human infant, highlights the biological and social importance of the extended developmental period of the brain and mind, involving the period of adolescence.

According to De Waal (2007), under communal conditions, gorillas and chimpanzees express political actions to obtain sexual, feeding, or hierarchical privileges that parallel human interactions to access dominance or political power, though in a modified conditioning scenario. In this context, animals' primary ways of dealing with disputes involve aggression with repressive behaviour (dominance), consensus, or submissiveness, which are also essential to human social settings. Based on social repression or 'socialization', cultural strata of variable 'thickness' have been constructed on top of drives implicit in our animal condition. However, these failed in deactivation, only being able to reformulate or condition their full expression. Under these circumstances, *friction* between these metaphorical 'geological plates'—of biological and cultural origin—will condition the emergence of various aggressive or maladaptive behaviours. Within this context, the *gap* between rational/instrumental knowledge and emotional updating would lead to the differential growth and expression of the emotional and rational components, or cognitive intelligence. The *friction* mentioned above tenses human individual and collective social construction, generating doubts about the true nature of a hypothetical 'post-human' future (Darlington, 2018).

The distant origin of the human species shifted from being integrated within interactive natural dynamics to one where human species advantage was the developmental shift of survival strategies through which the species became imperialists of the macrobiological world. Besides expressions of creativity in several domains, the human species progressively developed an advanced instrumental behaviour—leading to technological developments—coupled with a collective behaviour impregnated with mysticism, megalomania, an unyielding species pride, and fanaticisms deriving from pugnacious, belligerent fundamentalism and warmongering. These progressive collective developmental dynamics led the human species to the point of triggering beliefs that pretend to place their roots as marginal to the *natural kingdom* (Colombo, 2010). This trend evolved into a transition leading either to mysticism or *technocracy*, which tend to interfere with our integration with actual social, collective, dimensional, and personal profiles.

In this regard, perhaps one fundamental challenge facing the human species—besides its compromised survival due to uncontrolled technological, environmental degradation and weaponry development—would be to recognize its fundamental animal condition and aim to increase its *degrees of creative freedom* (Colombo, 2013) as well as cooperative social construction. Thus, it would be desirable that technological advances, potentially evolving into technocracy, and becoming a power factor, would not cancel or replace those subjective values and behaviours characterized under the concept of 'humanity'.

In general, theories about the evolution of *Homo sapiens* tend to coincide in that brain development correlates with the increasing complexity of its progressive social construction and the emergence of complex language formats. Social brain development was associated with the emergence of neural circuits involving prefrontal cortical regions and interactive neuronal ensembles coactivated by the processing related to norms and social rules implied in group living, reinforcing the road towards a complex social man, as demonstrated in Barbey et al. (2009). This would mean a correlation between processes that, in principle, do not share the same *tempo* in their evolutionary development—from the attributed emergency *Homo sapiens*—involving the cultural and neurobiological dimensions. At some point in this interactive context, the basic configuration of distributed neural circuits would have been expressed earlier in evolution. Thus, phylogenetically newer brain neural circuits would have undergone a continuous adaptive process to cope with the software organization's social and physical demands imposed through epigenetic reconfiguration during the last tens of thousands of years. This process involved the continuous generation/reformulation—mediated by neuroplasticity—of neural connectivity that conditioned the behavioural expression of persistent, ancient drives. In this regard, one of the 'engines' of the development of the human species' has been the exercise of its creative capacity and initiative that provided grounds for alternative behavioural expressions.

These considerations build a metaphoric view of biological and cultural interactions, conceptualized as interactive *tectonic plates*, as the engine fostering continuous mental development. In this regard, cultural banalization, poverty/indigence/marginalization of vast populations, social prejudices and privileges, and prevalence of financial, ideological, and religious fundamentalisms are at the base of unethical and counter-evolutive, regressive engines (Colombo 2010, 2013). They represent the instruments for domestication and suffocation of the species' search for freedom and creative potential, for the diverse interactive grounds on which brain connectivity and mental development ensue, coupled with phenotypic profile diversity, provide the primary means for the potential richness of the human species.

Universal animal drives, such as dominance/prevalence, survival, kinship, and 'profit' (greed, advantage, whether of material or social nature), provide the basis for the *evolutionary trap* promoting unstable, conflictive, dominant-prone human behaviours, both individual and group. The constant, dynamic 'trial and error' that characterizes the evolutionary process—benefitting the opportunities for the fittest for specific conditions of time and space—has led to the development of a progressively complex architecture of the mind.

In an interactive context, each species of the natural kingdom represents a relatively plastic substrate conditioned by the specific expression of its genetic heritage. It is apt to provide—within the constraints of its genotype—adaptive solutions for the survival demands of its natural niche and its community. Hence, some conditions reduce their degrees of freedom, which sets up a behavioural phenotype that tends to meet basic reproductive and survival demands. Coupled with a dynamic environment, it results in an unstable internal state of conflict/tension resolution. The final output of this interaction would result in consummation—achievement—or failure with the reprogramming of behaviour, or the replacement of the objective that reduces or eliminates such conflict/tension condition, or eventually in its extinction either as an individual or as a species.

The above considerations pose a conceptual restraint to our concept of freedom, which usually idealizes our natural capabilities, tending to ignore human limitations that arise from human ancestral animal origin and biological construction and its individual sociocultural history. Hence, the everyday concept of 'freedom' would instead build a metaphor, the literary construction of a utopia, to which normal use has attributed the character of a supposed reality. An integrated vision of the neural processes implicit in behaviour evolved in part with the evolution of the *social man*, precisely concerning the neurobiological and cultural dimensions. This would imply a correlation between processes that, in principle, do not share the same *tempo* in their evolutionary development during the estimated 200,000 years from the attributed emergence of *Homo sapiens*. As commented, the previous statement implies that at some point in evolution, the primary neural configuration emerged from an interactive context with the demands imposed by the progressively dynamic social and physical requirements. This implied a process of epigenetic reconfiguration of specific neural circuits—its software organization—during the last tens of thousands of years.

In other words, functional characteristics of the brain would not depend solely on the connectivity of specific regions/cortical areas and neuronal nets but also on the interactive, dynamic, generation/reformulation—neuroplasticity and behavioural plasticity—of local and expanded neural circuits dealing with inner and external environmental demands and interactions. In this regard, one of the 'engines' of human species development is, without a doubt, the exercise of its creative capacity and initiative since it lives—on metaphorical grounds, the consequences of the interaction of two *plate tectonics, biological* and *cultural*—with different history, constructs and adaptive dynamics.

The above considerations pose a critical point on the diversity of cultural environments and its impact on developing and promoting creative capacity and initiative according to the variable *genius* of the human species. Environmental diversities involve physical and social contexts and the historical inertia of community organization. Thus, the concept of 'relative freedom' must be assumed within the reality of the biological and historical configuration of the individual and its community. In this context, the conditions of deprivation, poverty, and marginality—unacceptable social conditions—create an actual risk of social harm to the human brain and mind. Conditions under which the creative potential of a large segment of our world population is significantly hampered or blocked, thus providing *human flocks* suitable for the undergoing cultural dominance process.

## References

Barbey AK, Krueger F, Grafman J. 2009. An evolutionarily adaptive neural architecture for social reasoning. *Trends Neurosci* 32: 603–610. doi: 10.1016/j.tins.2009.09.001.

Blair RJ. 2004. The roles of orbital frontal cortex in the modulation of antisocial behavior. *Brain and Cognition* 55: 198–208.

Boehm C. 2012. Costs and benefits in hunter-gatherer punishment. *Behav Brain Sci* 35: 1920. doi: 10.1017/S0140525X11001403. PMID:22289308.

Cole SW. 2009. Social regulation of human gene expression. *Curr Dir Psychol Sci* 18: 132–137. doi: 10.1111/j.1467-8721.2009b.01623.x.

Colombo JA. 2010. *Somos la especie equivocada?* (*Are humans the mistaken species?*). Eudeba. https://tinyurl.com/lsdozicg.

Colombo JA. 2013. *Bajo libertad condicionada* (*Under conditioned freedom*). Imago Mundi.

Darlington K. 2018. *Sistemas de inteligencia artificial que gestionan emociones humanas.* Open Mind BBA. https://www.bbvaopenmind.com/tecnologia/inteligencia-artificial/sistemas-de-inteligencia-artificial-que-gestionan-emociones-humanas/.

De Waal F. 2007. *Chimpanzee politics.* Johns Hopkins University Press.

dos Santos M, West SA. 2018. The coevolution of cooperation and cognition in humans. *Proc R Soc B* 285: 20180723. http://dx.doi.org/10.1098/rspb.2018.0723.

Dunbar RIM. 1993. Coevolution of neocortical size, group size and language in humans. *Behav Brain Sci* 16: 681–735.

Furuichi T. 2011. Female contributions to the peaceful nature of bonobo society. *Evol Anthropol* 20: 131–142. doi: 10.1002/evan.20308.

Gobrogge KL, Liu Y, Jia X, Wang Z. 2007. Anterior hypothalamic neural activation and neurochemical associations with aggression in pair-bonded male prairie voles. *J Comp Neurol* 502: 1109–1122.

Gordon TP, Rose RM, Grady CL, Berstein IS. 1979. Effects of increased testosterone secretion on the behavior of adult male rhesus living in a social group. *Folia Primatol* (Basel) 32: 149–160.

Jablonka E, Lamb M. 2007. Précis of evolution in four dimensions. *Behav Brain Sci* 30: 353–365.

Rose RM, Gordon TP, Bernstein IS. 1972. Plasma testosterone levels in the male rhesus: Influences of sexual and social stimuli. *Science* 178: 643–645.

Siever LJ. 2008. Neurobiology of aggression and violence. *Am J Psychiatry* 165: 429–442.

Striedter GF. 2006. Précis of principles of brain evolution. *Behav Brain Sci* 29: 1–12.

# 8

# HUMAN NATURE IN PERSPECTIVE

In general, successful survival strategies would vary according to species characteristics and the demands placed by any given ecological niche. Hence, each species continuously interacts with its habitat, so survival equilibrium is usually reached under a dynamic condition; thus, adaptive processes are continuously in demand, generating phenotypes and behavioural adaptations. From a fragile and dynamic equilibrium in a long timeframe of 'trials and errors' of the various evolutive series of the *Homo* species, a newcomer progressively disregarded the limitations imposed by the habitat and natural processes and demonstrated the ability to possess an added value resulting in a shift from environmental adaptation and interspecies equilibrium to ecological and species' dominance. This involved the optimized potential to generate an ample spectrum of plastic or adaptive behavioural strategies and to be able to modify the natural habitat. *Homo sapiens* succeeded in establishing and adapting across diverse Earth latitudes, involving modification of physical and environmental conditions and generating diverse types of communitarian organizations and communication means. Human current sophisticated behaviour evolved based on an ancient heritage of basic animal behaviours that have been exercised during millennia and adapted and installed under diverse environmental and biological demands. The enhanced mental performance provided by an increasing brain complexity allowed the expansion of behavioural strategies that then allowed adaptation and successful interspecies competence. However, this interaction did not occur without environmental costs, and different survival means among human collectives.

During the process of evolution, structural, and dynamic changes in neurobiological scaffolding allowed the emergence and combination of main traits that could be considered to make a *human profile*, i.e., creativeness, cognitive expansion, artistic or extrapersonal expression, progressive toolmaking, rich verbal communication, and intercultural solidarity. In turn, these evolved to come *under siege* from other behavioural vectors associated with culturally camouflaged ancient animal primary drives of dominance, material accumulation, territorial prevalence, and competing and privileged (reproductive and food accessibility)

DOI: 10.4324/9781003571544-9

behaviours. In a sophisticated, culturally moulded species, these derived into deviant individual/group behaviours that projected dominance into the realms of plain cruelty (when dominance behaviour extends beyond survival strategies), social inequity, and emotionally and physically expressed fundamentalisms.

Thus, the human brain and behavioural plasticity processed ancient animal behavioural drives that merged into various cultural expressions. These, in turn, moulded slowly maturing and adapting brain and mental processes. Thus, emotional and cognitive domains resolved into material evolution, providing continuously evolving tools that fed basic, ancient animal competing drives camouflaged under cultural profiles.

The migratory epic of early *Homo* may have several causes based on fundamental demands related to climate, food availability, and community needs, in addition to conflictive interactions among groups of settlers. A set of basic, shared animal drives met these demands. They strategically expanded characteristics in *Homo*, progressively involving curiosity and exploratory behaviour, defensive material and mystical means, and expanded behavioural prevalence unavoidably yielding to dominance attempts. Dominant behaviour represents a universal profile expressed within its own and foreign tribe members and towards environmental demands, including all living species. Hence, at the core of modern human creative and exploratory profiles survives the original behavioural drives that led the human species to its current planetary dispersion potential and condition of prevalence in the natural kingdom and historical human dominant drive profiles (Colombo, 2024a, 2024b). Increasing knowledge has been applied to the well-being and oppression of human beings—often applying cruel means, a behaviour of exclusive human domain. In this regard, it seems reasonable to remark that programmed cruel behaviour—i.e., those terminal behaviours unrelated to the survival needs, such as programmed physical or emotional torture—emerged together with increased brain/mental capabilities, perhaps associated with prefrontal neural development and its downstream brain interactions.

The progressive emergence of humans with sophisticated language allowed the development of complex virtual constructs based on symbolism and the instruments of culture, which have enhanced cognitive capacity, creativity, and emotional interaction. These are processes anchored in neural networks distributed within interactive cortical and subcortical networks, still retaining essential elements and connectiveness preserved during evolution. They include the hedonic attractiveness of power or prevalence, securing territory, dominance, and drives towards reproductive and food rights. They became a current component of the evolving cultural process that involves competition, consumption, accumulation, violence/belligerence, personal contribution, narcissism, and privileges—all these being generators of inequity under hierarchical constructions.

At this evolutive stage, it should be considered that territoriality and prevalence are universal domains in survival behaviours of occasionally related and gregarious species within the natural kingdom. Hence, whether solitary or gregarious social habits, animals and vegetables (Baluska and Mancuso, 2020) show territorial competitive behaviours towards conspecifics and prey on them or reject those competing for common feeding sources and survival or reproductive resources. *Homo sapiens* carries such a universal backpack, and the expression of its culturally transformed—or hidden—animal drives take place as dominant behaviour and hierarchical social constructs, either spontaneously, under social dynamic circumstances, or converted into virtual (cultural, ideological) domains. It should be added that this mental/behavioural power grew with a bonded relationship with the continuously evolving sophisticated material culture. This material and technological bond conditioned the evolution of human social behaviour since the evolution of *Homo sapiens* became wrapped around the construction of instruments of progressive complexity and power and developed into a continuously evolving cultural and material technology that dynamically and progressively reset the relationships among individuals and between them and the environment. This projects onto current social trends. Sociocultural differences among worldwide communities and, within them, among their constituents result from a different history and dynamics of genomic-environmental interactive human constructions and justify limiting globalization's current general concept to limited socioeconomic domain strata.

Which and how many of human current behaviours—individually and as a global community—are driven by ancestral, inherited traits imprinted in the human animal condition suggests the need to re-evaluate current social drives in different cultures and sociopolitical systems. The cultural domain implied in this question is intimately connected to human identity and pertinence to social constructions and environmental/ecological invasive degrees driven by profit or survival convenience.

The individual and collective behavioural span involves a series of intervening neurocognitive executive levels. Yet not all events span the conscious dimension. Cognitive processing involves distributed neural circuits as a processing substrate. From an intellectual point of view, the most intriguing event is that much of the former appears to be at the subconscious level. According to some authors, what emerges at the conscious level, expressed temporarily at a specific time, would be events we can manipulate as working memory in our executive behaviour domain, corresponding to the *explicit memory* (Dietrich, 2015).

Animal nature involving animal drives has been progressively built and placed into practice since ancient times—based on survival, social gain, and profit—and is compressed to our phylogenetically basic neural systems and basic survival behavioural construction. It plays a role in human social interactions and cultural constructions (Colombo 2019, 2021a, 2021b) and plastically adapts to multiple behavioural demands. No wonder the human species has been considered to have a bipolar behavioural profile (Boehm 2012a, 2012b) attributable to genetic lineages spanning from the ancestral *Pan* and schematically profiled by the behaviours of chimpanzees (tendency to conflict, male predominance) and bonobos (preventive behaviour, female predominance). Besides the cultural environment and values, each ethnos has interactively developed. This anticipates a potential behavioural bipolarity with an uneven representation and prevalence distribution among individuals and collectively on social organizations' profiles.

Through time, cultural strata of variable composition have been developed on top of drives implicit to our animal condition, based on social repression or 'socialization'. Nevertheless, they failed in the deactivation or suppression of those ancient drives and only succeeded in diverting (camouflaging) their expression or reformulating or transitorily repressing them—thus increasing pressure on repressed basic drives—as the history of human civilization demonstrates, as well as conditioning deviant behaviours, expressed at the individual and collective levels. Interaction with the physical and cultural environments continues shaping our ethnic behavioural variations. Yet, our primary organization is bound to ancestral demands imprinted as a set of basic drives (territorialism, reproduction, survival, secure feeding sources, dominance, and cumulative behaviour). Their expression, affected by changed environmental (physical and sociocultural) conditions, sets up the probability of continuous friction between the neurobiological and cultural *tectonic plates* (Colombo 2010, 2015).

The plastic, interactive nature of the human brain and mind construction (depending on cultural issues and contexts) provides for adaptative behavioural responses. However, these have remained the framework of ancient primary drives (as mentioned above) imprinted in the heart of *Homo* animal construction that affect the sociocultural profile or the probability of their expression. That is, biosocial interactions continue to model sociobehavioural trends, or social phenotypes, on top of the primary, deeply entrenched survival and prevalent drives conformed according to the basic structure of the ancestral animal nature. This has conditioned the disparate social, cultural, and cognitive conditions among nations, ethnicities, and individuals that have contributed to build the crackled composition of our modern world. They include distribution of economic resources, political and financial dominance, unequal rights, poverty, disparate unequal financial wealth, relative access to cognitive development and quality of life, and individual and collective responses to those circumstances.

Primitive or basic *Homo* interactive constructions were expressed in the original ancestral communities based on standard parameters common to other mammalian species: hierarchies, group priorities, survival drive, territorialism, food search, and reproductive strategies. Yet, progressively, the cultural complexity and sophistication of those basic tenets resulted in collective 'moral sets' and their institutionalization, which replaced the herd interactions that conditioned individual behaviours. These were incorporated into the cultural heritage, as language and symbolic structures enriched communication and complex social organizations—subtle changes in brain organization mediated through neuroplasticity processes made such development possible. It included adaptations to different physical and social environmental conditions that shape biosocial interactions.

Such a highly dynamic process continued to the point of cognitive development allowing virtual constructions to be incorporated into its perception of *reality* and beliefs. This interactive process slowly allowed the construction of a virtual parallel dimension to the *natural history* that pervaded human cognition, foreign to its natural origin and sometimes fed by hallucinations and behavioural disorders. Its strength merged from constructing an imaginary dimension where supernatural beliefs offered a sense of security and expectations during a human lifetime and projected beyond death. It also created a moral framework to build human relationships and resolve the notion of fragility, fear, guilt, and deep uncertainty.

If not manifest, basic profiles of human-animal origin remained occult, repressed, or rejected in such parabiological constructions that occasionally reached the condition of fundamental beliefs. As human history shows, these were able to display various worrying levels of vicious hostility and cruelty, curtail human freedom of thought, and share natural material inequities. In attempting to 'shake off' some restraining chains shared during human evolution under variable cultural and ecological domains, human civilization fell prisoner to deeply entrenched emotional belief constructions.

The human species' knowledge, creativity, and technological developments generated a dominant macro species that, at the same time, tried to avoid attaining full consciousness of its animal origin and fundamental biological nature. This condition affects some of its emotional expressions and social drives (territorialism, reproductive and feeding priorities, survival, and prevalence). The aggressiveness, cruelties, social inequities, and unrelenting individual and socioeconomic class ambitions are the best evidence that humans must first recognize and assume their fundamental nature to change their ancestral drive. Profound cultural changes are only possible and enduring if humans come to grips with their actual primary condition.

To what extent does our animal condition affect our individual and social behaviours? To what extent is such a primary condition considered in constructing ideologies, social and political systems, and policies? Do pedagogic strategies include some consideration of human natural construction? In fact, can they regenerate human individual and social values without considering it?

## References

Baluska F, Mancuso S. 2020. Plants, climate and humans: Plant intelligence changes everything. *EMBO Rep* 21(3):e50109. doi: 10.15252/embr.202050109.

Boehm C. 2012a. Costs and benefits in hunter-gatherer punishment. *Behav Brain Sci* 35: 1920. doi: 10.1017/S0140525X11001403. PMID:22289308.

Boehm C. 2012b. Ancestral hierarchy and conflict. *Science* 336: 844–847. doi: 10.1126/science.1219961. PMID: 22605759.

Colombo JA. 2010. *Somos la especie equivocada?* (*Are humans the mistaken species?*). Eudeba. https://tinyurl.com/lsdozicg.

Colombo JA. 2015. *Los Homo sabios. Globalizados o segmentados?* (*The wise Homo. Globalized or segmented?*) Buenos Aires Books.

Colombo JA. 2019. *Our animal condition and social construction.* Nova Science Publishers. ISBN: 978-1-53615-357-6. eBook ISBN: 978-1-53615-3583.

Colombo JA. 2021a. *The homo within the sapiens.* Nova Science Publishers. ISBN 978-1-53618-373-3. https://tinyurl.com/3wbdchlc.

Colombo JA. 2021b. Ancestral drives and the complex social behavior of human species. *Int J of Primatol Res* 4(1): 18–19. https://www.researchgate.net/publication/349768145.

Colombo JA. 2024a. *Evolution and the human-animal drive to conflict. A psychobiological perspective.* Routledge.

Colombo JA. 2024b. *Human biome and human behaviour.* Routledge.

Dietrich A. 2015. *How creativity happens in the Brain.* Palgrave-Macmillan. doi: 10.1057/9781137501806.

# 9

# DOMINANCE IN EVOLUTION

> Today, humans retain the genetic underpinnings of both conflict and conflict management; thus, we retain the potential for both war and peace.
>
> (Boehm, 2012)

As stated by Boehm (2012), ancestral traits are identifiable in the species derived from the ancestral *Pan* (*Pan troglodytes*, *Pan paniscus*, *Homo*), though they are also shared by other, non-primate animal species, with variable social profiles. Those traits include social binding, conspecific conflicts, territorial dominance, hierarchies, and behavioural profiles involving group size dominance, hierarchies, and submission. These behavioural profiles promote both fight engagement and peace arrangements within social dynamics.

The emergence of progressively sophisticated toolmaking provided the basis for a progressively increased physical and economic engagement applied to social dominance that evolved into starch inequalities and global conflicts among humans. Thus, the ancestral drive towards survival derived into various dimensions of social dominance and expanded conflicts, potentiated by increased knowledge and technological developments in the human species.

Yet, at the base of this progressive development lie the simple ancestral biological forms that survived in parallel to later developed, multicellular complex organisms. The existence of more biologically complex beings emerged from elementary primary forms of life. In this context, leaving aside precisely those initial forms of life, of all the taxonomic groups of the natural kingdom, insects—which would have originated about 390–350 million years ago (Nel et al., 2013)—are probably the most abundant today, constituting approximately three-quarters of the recorded fauna. Given the persistence over time of this group, it could be thought that its evolutionary dynamics represent one of the most effective examples of biological dominance. Our coexistence with this group often acquires the characteristics of a defensive strategy considering their role as human disease vectors—e.g., *Triatoma infestans*, mosquitoes, flies, parasites, and bacteria. Often, human resources attempting to

DOI: 10.4324/9781003571544-10

continue exercising its *macrobiological imperialism* collide with microscopic and ubiquitous beings, no matter what the degree of attempted control, using antibiotics, vaccines, and antiseptics. Some forms of life have survived in the harsher, more extreme living conditions than we now know. These are the so-called 'extremophile' microorganisms, either because they survive in acidic (acidophilic) or alkaline environments (alkaliphiles), under high pressures (barophiles), or in rocks or deep soil (up to 2,700 m depth), or with radiation (radiophiles), in high temperatures (close to 80 °C) (thermophiles or hyperthermophiles), or without water.

The possible 'evolutive path' species of the genre *Homo*, including *sapiens*, generated anthropological hypotheses involving the possibility that our direct ancestor *Homo* would have been the *Homo erectus*, or variations that emerged from it, such as the *Homo antecessor* and *Homo heidelbergensis*. The *Homo erectus* would have made successive migrations within Africa as well as towards the Euro-Asiatic continent, leaving enclaves that were later invaded in different regions by other *Homo* migrant species, including the *anatomically modern sapiens* and Neanderthals. Different phenotypes and ethnicities would have evolved from successive migrations, interbreeding, and ecological adaptations. The possibility that a variety of hominids preceded the human species, some of them occasionally cohabiting with the modern *sapiens*, thus forming a large collective, was reported in an editorial commentary in the journal *Nature* (2016). Theories regarding the hypothetical evolutive chain leading to the human species may vary, and different routes and possible genetic exchanges may be recognized, as Kuhlwilm et al. (2016) consider.

Possible stages in the evolutive process of the *hominids* involving transitions due to reductions in the demographic density of various lineages were considered by Foley et al. (2016) and Lahr (2016). Thus, changes in hominid demographic distribution could have promoted the emergence or differential expansion of *Homo* lineages, generating transient stages in the evolutive display. The agreement seems to converge into the concept that the origin of *Homo sapiens* would have emerged following a succession of hominid species, each one of them characterized by structural components and functional and survival strategies, as discussed by the above-quoted authors. Thus, our closest ancestors would have evolved from a variety of *Homo* lineages with differential interbreeding, added to epigenetic variables that completed the speciation of the modern human with its phenotypic diversity.

Due to gregarious conditions, social organization patterns involving relationships between individuals, the group structure, and the environmental ecosystem would have differentially modified the neurobiological substrate supporting individual behaviour. Thus, increased social demands placed on the neurobiological organization of the genre *Homo* would have progressively resulted in the *brain incarnation* of culture—a comparatively more straightforward process involving individual and group interactions among other mammalian species. These demands—coupled with its progressive ecological dominance—were progressively based on values, beliefs, language, symbolic constructs, productive chain, neoteny of the newborn, coupled with other survival conditions (feeding habits, social organization, sedentarism, reduction of the stress evoked by risks due to survival as a prey) that promoted microstructural changes affecting neural circuits associated to social organization. Such progressively increasing demands in *Homo* would exceed in complexity the basic rules/codes present in non-human gregarious animal communities with a hierarchical organization. As discussed in Renfrew (2008), such cultural development would have had either an extended, slow, progressive profile or an accelerated one—the so-called *cultural revolution*—during the last 40,000 years according to others, probably associated with the emergence of language. According to Mcbrearty and Brooks (2000), the time lag before the appearance of the *Homo sapiens sensu strictu* dated more than 100,000 years, while the claimed emergence of modern human behaviours known as the *human revolution* would have taken place ca. 40,000–50,000 years ago. This would support the notion that the earliest modern humans were behaviourally primitive. According to Renfrew (2006, 2008; Renfrew et al., 2008), this apparent uncoupling between the genetic, physical, and cultural configuration of the *anatomically modern Homo sapiens* would represent the *sapiens paradox*.

What neurocognitive correlations must have taken place to develop a symbolic expression? The temporal coincidence of parietal graphics and artefactual expressions and evidence of ritual behaviours would suggest that they would have evolved in an interactive context involving distributed neural circuits, including the phenomenon of *exaptation*—recruitment of neural circuits or structures associated with specific functions to be involved in other functional domains—and involving new connectivity. According to Tattersall (2004, 2010, 2017):

> The extraordinary cultural achievements of the Upper Paleolithic Europeans show that by 40 kyr ago the exploitation of the human symbolic potential was well advanced; but wherever and whenever this highly generalized capacity may have been initially discovered, substantially after the birth of *Homo sapiens* as an anatomical entity, the discovery of how to use it in new ways was a gradual process and one that is still proceeding.
>
> *(Tattersall, 2004)*

> Nothing indicates convincingly that any hominid behaved symbolically prior to *Homo sapiens*, which originated 200,000 years ago but started behaving symbolically only 100 k yr. later.
>
> *(Tattersall, 2017)*

In this regard, it is interesting that Adams (1998) developed the concept of *synaptic Darwinism*, according to which neural connections are cancelled or reinforced or generate conditional new ones based on the concept of the rate of activity of neural circuits. This proposal implies the combination of concepts based on Darwin's theory applied to species evolution and on the dynamic synaptic concept of Hebb (1949) applied to synaptogenesis, synaptic lysis, and synaptic reinforcement at the microstructural level.

The human imperialist species represents the construction of a biological empire where its only predators—besides microorganisms—are its own members.

Opinion-makers have synthesized today's world in which humans live by coining *globalization* to define a stronger link between economic events and information. This misleading term tends to replace the use of the word *humanity* and, in doing so, has redefined the interactions of the world community irrespective of any human inclusion. It argues that globalization is an economic, technological, social, and cultural process on a grand scale, with increased communication and interdependence among countries worldwide, achieved by unifying their markets, societies, and mainstream cultures. Yet, the concept seems to disregard that such an intercommunication trend has been in progress since humans almost completed invading planetary niches. Their respective populations grew, affected by improving communication means and, even more so, language has probably been a central cause of the rapid evolution of human civilization.

Conquest and colonization, as an expression of dominance drives, have been among us since the dawn of *Homo sapiens*, whether to achieve more significant revenues or energy resources or to satisfy expansionist tendencies—territorial, access to resources, cultural, and financial. But, more important yet, such a concept seems to disregard the asymmetries that profoundly divide the living conditions of communities and individual citizens within them. In simpler words, the human world does not operate as an *equal opportunity* employer, an *equal opportunity* educator, or provide conditions for an *equal opportunity* citizenship.

Let us recognize and include the importance of the speed of information used nowadays in decision-making. However, we should also realize that spy satellites and the non-proliferation of nuclear weapons are issues solely controlled by a small group of superpowers that supposedly struggle to maintain feeble 'peace' conditions worldwide or protect their regional interests and prevailing conditions. Around them spin the rest of the world's communities, aware of the inequity but often unable (or prevented by economic or political means) to initiate a change. The proposed thesis is that the so-called *globalization* runs the risk of deepening the *segmentation of humanity* in terms of quality of life, socioeconomic-military dominance, and educational level. When the rate of social, scientific, and technological change is not uniform but, rather, significantly asymmetric, what tends to happen is that, as in any competitive environment, dissimilar groups begin to form. The front-runners set the rhythm and the developmental pace. The rest are either followers or are excluded and become marginals or dropouts, or decide to opt for other directions, which may eventually come into conflict.

Knowledge must be allowed to flow. However, the long-term strong asymmetries among the countries' economies, education profiles, productivity, and predominant conditions or trends in territorial, commercial, and cultural prevalence or dominance, turn hollow the humanistic concept of globalization. Thus, it becomes only understandable in terms of accessing information by limited groups of individuals within communities or groups of nations (world statistics on internet penetration and other measures of life conditions are clear indicators of such asymmetries, rather than 'global integration').

Developmental conditions are extremely uneven among countries and world regions regarding the information offered, amount, relative significance and objective quality, demand, public interest, or interpretation skills. Within them, there is also the potential condition of the individual receptor to transform such information into personal development or informed positioning. The information reaches a comparatively small segment of the world population through traditional means or the *web*. In addition, as said above, the ability to read, translate, understand its spin-offs, and respond to such information is affected by the sociocultural conditions of the receptor, which includes preparation to search and evaluate meaningful information for comparative data analysis and context awareness. In other words, information only reaches some communities and—within them—a minority with physical and functional (cognitive) access to media.

Besides the potential to generate new links, bonds, interests, or sensibilities in well-defined segments of the world population, the trustworthiness and quality of information—and its manipulative potential—are questionable. As we witness currently, these minority communities could mobilize other sectors of their community. But it is still basically a minority that manages such a level of information, a minority that is 'synchronized' by similar social codes or requirements, financial codes, needs, or global demands.

It appears as if people live in imaginary cultural, economic, and political 'bubbles', from which they may or may not communicate with others. Still, they can only access some of them, i.e., those that express some common denominator(s). Based on their 'bubble pertinence', each person builds their vision of 'reality'. Those who are 'globalized' cohabitate with a particular set of 'bubbles'. The level of social awareness defines a main characteristic of each 'bubble', i.e., those who live on less than 1 or 2 US dollars per day, those without primary care, or those who have no opportunity to access education or access to the internet, or who were not raised in enriched environments that foster optimal development, or those who had access to critical levels of education. However, this 'connectivity' provided by modern communication technologies does not glue together social 'bubbles' in which needs and access to goods and services, interpersonal or community codes, and priorities may differ dramatically.

Let us go back to the notion that communities—or nations—on the planet are not organized in egalitarian terms due to natural, historical, or political reasons. Further yet, most communities are inherently non-egalitarian despite presumed, iteratively declared, democratic political structures. Also, let us remember that the notion of dominance is central to the relationships between most nations. Hence, *globalization* does not necessarily universally promote a better quality of life and rather may promote further segmentation—further inequalities. If we consider the media organizations that manipulate communication and have the possibility of modulating people's opinions, we would be tracing the configuration of strategies of multinational monopolies or oligopolies or the identity of the trade policy interests of these organizations and relationships with governments, with who eventually merge in political and economic aims and strategies. The meshwork among political and corporate interests does not warrant the flow of trustworthy public information. In this regard, it may suffice to take note of the predominance of the nuclear weapons power of certain nations, the creation of an international controller/inspection body composed of the atomic power nations themselves to limit the proliferation of such weapons by other countries, or the incredible creation of a space station at the cost of around 100 billion euros (*Bulletin European Space Agency*), or the costs of the arms race, to confirm these asymmetries that make the concept of *globalization* more a matter of selling a *cliché* than of real life.

In summary, there can be no globalization in a real sense in a profoundly segmented—asymmetrical—planetary community where the flow of meaningful information occurs within specific *social niches* or *strata*. And, where the question remains regarding the credibility of public information, it became a powerful tool to condition public opinion rather than an instrument for humanity integration. It may serve to interact with stock markets and transactions between companies, to run financial business among them, or to mobilize specific social segments aiming at political or commercial targets. But always within a given set of 'bubbles'. This essay stresses the profound distortion in terms of quality of living, access to information, and opportunities for social mobility that calls for a significant revision of our social values and priorities.

In this panorama, it is relevant to consider some aspects of the so-called 'globalization' process. This consideration should not leave aside concepts of implicit relative Darwinism, where the evidence of inequality in the distribution of wealth and the ambitions of financial operations will drive the path towards the deepening of the misery of the most neglected sectors.

## Globalization and Dominance

One of the questions that—in different forms—is present throughout these comments is whether a 'mosaic' of social behaviours may exist, probably linked to the genome and the development of neural circuits expressing greater or lesser ability—whether biased towards submissive or aggressive social behaviours. In this regard, whether such ability constitutes a substrate for cultural influences and the construction of the phenotype should be considered. This could constitute both an individual and group (ethnic?) '*marker*', that is, expressed occasionally or as a gregarious characteristic, depending on the developmental conditions and the uniformity of the genetic pool.

Interestingly, this notion of 'mosaic' has been incorporated as an intrinsic construction of the evolutionary process. As a case of mosaicism, it is worth mentioning the fossil specimen of *Australopithecus sediba*, analysed by Kivell et al. (2011), concerning the anthropometric study—and its extension to possible body dynamics—of the evolution of the immediate ancestors of the *Homo* genus. Kivell et al. (2011) underline that comparisons to other fossil hominins suggest at least two distinct hand morphotypes around the Plio-Pleistocene transition. In this regard, *Australopithecus sediba* may represent a basal condition associated with early stone tool use and production. In general, the concept of 'mosaic' has been used in the field of evolution concerning body components or segments, a matrix on which evolutionary pressure would have acted to generate variants in the natural kingdom. For example, neural circuits with nodes anchored in specific brain structures (such as the amygdala, in the temporal lobe of the brain, and areas of the prefrontal cortex and the cingulate) have been associated with the development of social complexity that characterizes *Homo sapiens* (see Dunbar, 2003).

If such a 'mosaic' exists—in what refers to 'innate' brain components linked to basic tendencies of social behaviour or modified from early ages of ontogenetic development—the need to reinforce the assignment to cultural and educational domains should be considered. That is, whether they promote prosocial and creative behaviours due to their central role in social construction. The components of the selfish behaviour of the human species must have convergent correlates from the neurobiological and the cultural domains, able to be modulated and enriched. Its projection to the generation of new leaders and the modification of the parameters of social behaviour constitute a notable aspiration in the relatively recent path of the human species. Added to these terms of our vital equation, the dynamics of today's trends will determine the possibility of generating a 'humanity' enriched—instead of threatened—by cultural variety. For this reason, 'globalization' has risen due to oversizing the scope of an instrument—communication—rather than of humanity integrated and revalued by that instrument.

The thesis presented in this chapter is concerned with the elements that drive the so-called 'globalization' to run the risk of deepening the segmentation of humanity. As long as the organization of the social fabric is ***not*** firmly anchored on criteria of solidarity, equity, individual and collective responsibility, and educational level, it will increase social segmentation. Additionally, they could thus either generate the extinction of cultural identities or feed the conditions for confrontations. As long as the speed of change in knowledge and scientific-technological application accelerates without correcting sociocultural asymmetries, individual and collective development conditions will keep feeding its progressive social distortion.

The following statement by Romanello et al. (2023) summarizes the impact of current country developmental profiles on health (and educational) management:

> Only 44% of low Human Development Index countries and 54% of medium HDI countries reported high implementation of health emergency management capacities in 2022, compared with 85% of very high HDI countries.
>
> *(p. 2347)*

Analysis of species survival behaviour depicts a wide range of strategies adapted to the interaction of individuals/groups with the environment and genetic/phenotypical characteristics of the species and individuals. Survival rates depend on such interaction, where the main differential living species profile with humans would reside in the range and behavioural configuration of fundamental drives.

Primary goals refer to survival (defence of feeding territory) and reproduction, involving competence, dominance, and sociability. Through millennia, these behavioural profiles and vectors provided the range of species and individual behaviours that fed natural interactions. Human species evolved, expanding this behavioural range by adding additional domains that enlarged competence arenas and territorial dominance (Colombo, 2010, 2021a, 2021b). These are well beyond basic survival needs where material luxury, weapons of massive destruction aimed at enlarging or protecting territorial and economic prevalence, outplaced human creativity in solidarity and aesthetic domains. As an example of current world spending priorities, the following brief from the report by the Stockholm International Peace Research Institute (SIPRI)[1] states that:

> Global military expenditure rose for the eighth consecutive year in 2022 to reach an estimated $2240 billion, the highest level ever recorded by SIPRI.
>
> *(p. 1)*

Though a minority of world communities managed to emerge from the dominant behavioural trends, an unknown but probably considerable community of creatives would remain obliterated due to conditions involving personal development, economic prejudice, and lack of opportunity to express their potential as social modifiers. Additionally, for those who manage to survive creatively, their creations are accessible and understood by only a minority of the world population due to educational and socioeconomic reasons. Thus, the social impact of the creativity dimension remains sequestered and restricted due to current educational and economic stratification.

The following excerpts construct a critical perspective of current economic and educational imbalances.

According to the *World Bank* report (2023), almost 700 million people worldwide live in extreme poverty, a subsistence of less than $2.15 per day, involving a large percentage from sub-Saharan Africa.[2]

These figures are supported by a *Statista* (2023) report.[3]

According to *Oxfam International* (2024),[4] this contrasts with the fact that Aliko Dangote—Africa's richest person at the time of that report—holds a 'near-monopoly' on cement in Nigeria. His empire's expansion into oil has raised concerns about a new private monopoly. Additionally, according to this source, since 2020, the wealth of the five richest men has doubled, while 5 billion people were made poorer in the previous decade.

According to the 2022 report by the *International Monetary Fund* (A Stanley, 'Global inequalities'), the richest 10 per cent worldwide today snatch 52 per cent of world income, while the poorest 50 per cent represent only 8.5 per cent of the global income.[5]

The report from *Developing Initiatives*[6] further reinforces this picture, stating:

> Estimates for 2022 suggest that 46 million more people were living on less than the $3.65 poverty line than in 2019. As of 2022, 23% of the global population (1.8 billion people) were living on less than this threshold … In 2022, we estimate that 1.85 billion people (26% of the global population) lived below the threshold of $3.65 a day and 3.71 billion (46% of the global population) lived below the threshold of $6.85 a day. Thirteen of the 15 countries with the largest increase in the number of people in extreme poverty are in sub Saharan Africa.

Monetary poverty represents one factor considered a condition of 'multidimensional poverty', which also involves other dimensions, such as monetary poverty, education, and basic infrastructure services. In such regard, as reported by the *World Bank* (2022):[7]

> While monetary poverty is strongly correlated with deprivations in other domains, this correlation is far from perfect. The Poverty and Shared Prosperity 2022 report … shows that almost 4 out of 10 multidimensionally poor individuals (39 percent) are not captured by monetary poverty, as they are deprived in nonmonetary dimensions alone.

The following examples are extracted from the *Multidimensional Poverty Index* (MPI):

- In Mauritania (2023), 2.3 million people, or 56.9 per cent of the population, live in multidimensional poverty. These people are deprived on average in 56.3 per cent of the weighted indicators in education, health, living standards, and employment.[8]
- In Sri Lanka (2023), approximately one out of every six people are multidimensionally poor, and in estate areas, there are pockets in which 51.3 per cent of all people in these areas are living in poverty.[9]
- In India (2021), multidimensional poor (based on national MPI) amounts to 25.1 per cent.[10]
- In Belize (2021), the results of the MPI study revealed that 35.7 per cent of the population lived in multidimensionally poor households.[11]
- In Paraguay (2021), 24.9 per cent of the population lived in multidimensional poverty.[12]
- In Guatemala (2019) 61.6 per cent of the people lived in multidimensional poverty.[13]
- In Uganda (2022), multidimensional poor (based on national MPI) reached 42.1 per cent.[14]
- In Afghanistan (2019), the multidimensional poor reached 51.7 per cent of the population.[15]

In 2023, the *International Journal of Rural Development* released its analysis of trends from 2000 to 2022. This followed the latest update of the global *Multidimensional Poverty Index*, with estimates for 110 countries released by the United Nations Development Programme (UNDP) and the Oxford Poverty and Human Development Initiative (OPHI) at the University of Oxford, UK, in July 2023. The *International Journal of Rural Development* stated:[16]

> 1.1 billion out of 6.1 billion people (just over 18 per cent) live in acute multidimensional poverty across 110 countries. Sub-Saharan Africa (534 million) and South Asia (389 million) are home to approximately five out of every six poor people.

It ought to be stressed that the above-reported data represents annual updates, which conceals the cumulative factor of years of inadequate nutrition and education; thus, the cumulated affected population numbers should significantly exceed the annual updates of individuals existing under deprived or developmentally delayed conditions, as well as deepening their sociocultural condition with time. This implies an ever-growing population with developmental opportunities obliterated or significantly reduced by their poverty and educational conditions as well as by the effects on cognitive development and health care conditions. Thus, current statistical numbers represent a partial view, for they represent a snapshot of a long-term, cumulative degrading dynamic process that has an accumulated profound impact on neurobiological and social development affecting individual and collective growth and rights, with unequal world distribution.

As stated in a *Nature* editorial (2016):

> Were we able to mark the remains of all our ancestors from that point, the world would be an enormous cemetery.

It added:

> ... it would provide yet more evidence that the human lineage is more diverse than ever imagined.

The racial and ethnic variety, which emerged as a product of the adaptation of *sapiens* to very dissimilar ecological niches and adaptable variations of the genetic menu, provided different adaptive alternatives and cultural richness. A cultural variety that today is menaced, in addition to the risk of survival that other biological species face today. Multiculturalism, as an element of an ideology, goes hand in hand with the defence of biodiversity. Both are equidistant from the objective of equity. However, it should be considered that scientific-technological progress has maximized thought trends that prioritize the species' exploratory and creative nature. In this sense, in the human horizon, for example, the unrestricted development of extraterrestrial targets has been proposed as an objective. This includes the possibility of generating new phenotypes from adaptation to extraterrestrial planetary conditions and manipulating the human genome (Smith, 2013), given the near possibility of achieving transgenic primates (Shen, 2013).

Regarding the formidable and complex ecological impact of human civilization on the planet, it should be mentioned that geologists and palaeoanthropologists are already considering incorporating the *Anthropocene* era—a term coined a decade ago by the Nobel Prize winner in Chemistry, Paul Crutzen—into the classification of geological eras, in addition to anticipating a new era of species extinction (see discussion in Leakey and Lewin, 1995; Mora et al., 2011).

It is impossible to continue ignoring all these ominous signs of human actions and their impact on their fellow human beings and the planet's environment. It ought to be stressed that the *species' adolescence* appears endless.

The formidable challenge for the future—long overdue—lies in finding the formula for developing social and biological equity policies for life on this planet without blocking some of the human species' fundamental characteristics: its exploratory vocation, creativity, and ingenuity, with no limit placed on the acquisition of knowledge.

Considering that the cultural process is mounted on primary biological and social contexts, such as cognitive enhancement access, brain neuroplasticity, and heritable epigenetic changes, the abysmal differences in social goals and complexity of interpersonal relationships among communities should not come as a surprise. This sets up the basis for divergent futures among human populations, with its extremes represented by cognitive and technological development-prone and relegated, cognitively underdeveloped, impoverished individuals and communities, as well as diverse social constructions. The progressive impact of access to complex sources of knowledge and technological know-how has maximized historical divergence among communities.

## Notes

1 Global developments in military expenditure, 2022. https://www.sipri.org/yearbook/2023/05.
2 https://www.worldbank.org/en/topic/poverty?
3 https://www.statista.com/topics/781/poverty/#topicOverview.
4 https://www.oxfam.org/.
5 https://www.imf.org/en/Publications/fandd/issues/2022/03/Global-inequalities-Stanley.
6 https://devinit.org/resources/poverty-trends-global-regional-and-national.
7 https://www.worldbank.org/en/topic/poverty/brief/multidimensional-poverty-measure.
8 https://www.mppn.org/mauritania-launches-multidimensional-poverty-index/.
9 https://www.mppn.org/sri-lanka-launches-official-national-mpi-linked-to-a-child-mpi/.
10 https://www.mppn.org/paises_participantes/india/.
11 https://www.mppn.org/multidimensional-poverty-index-to-complement-monetarypoverty-measurement-in-belize/.
12 https://www.mppn.org/paises_participantes/paraguay/.
13 https://www.mppn.org/paises_participantes/guatemala/.
14 https://www.mppn.org/paises_participantes/uganda/.
15 https://www.google.com/search?q=AFGHANISTAN.
16 https://www.rural21.com/english/news/detail/article/the-state-of-multidimensionalpoverty.htm.

## References

Adams PJ. 1998. Hebb and Darwin. *J Theor Biol* 195: 419–438.

Boehm C. 2012. Ancestral hierarchy and conflict. *Science* 336: 844–847. doi: 10.1126/science.1219961. PMID: 22605759.

Colombo JA. 2010. *Somos la especie equivocada?* (*Are humans the mistaken species?*), Eudeba. https://tinyurl.com/lsdozicg.

Colombo JA. 2021a. Ancestral drives and the complex social behavior of human species. *Int J Primatol Res.* https://www.researchgate.net/publication/349768145.

Colombo JA. 2021b. *The homo within the sapiens.* Nova Science Publishers. ISBN 978-1-53618-373-3. https://tinyurl.com/3wbdchlc.

Dunbar RIM. 2003. The social brain: Mind, language, and society in evolutionary perspective. *Ann Rev Anthropol* 32: 163–181.

Foley R, Martin L, Lahr MM, Stringer C. 2016. Major transitions in human evolution. *Philos Trans R Soc Lond B* 371: 20150229. http://dx.doi.org/10.1098/rstb.2015.0229.

Hebb DO. 1949. *The organization of behavior. A neuropsychological theory.* Wiley.

Kivell TL, Kibii JM, Chuchill SE, Schmid P, Berger LR. 2011. *Australopithecus sediba* hand demonstrates mosaic evolution of locomotor and manipulative abilities. *Science* 333: 1411–1417.

Kuhlwilm M, Gronau I, Hubisz MJ, Filippo C, et al. 2016. Ancient gene flow from early modern humans into Eastern Neanderthals. *Nature* 530: 429–433. doi: 10.1038/nature16544.

Lahr MM. 2016. The shaping of human diversity: Filters, boundaries and transitions. *Philos Trans R Soc B* 371: 1–12. http://dx.doi.org/10.1098/rstb.2015.0241.

Leakey RE, Lewin R. 1995. *La sexta extinción* (*The sixth extinction*). Tusquets.

Mcbrearty S, Brooks AS. 2000. The revolution that wasn't: A new interpretation of the origin of modern human behavior. *J Human Evol* 39: 453–563.

Mora C., Tittensor D.P., Adl S., Simpson A.G., Worm B. 2011. How many species are there on earth and in the ocean? *PloS Biol* 9(8): e1001127.
Nature. 2016. Editorial: Humanity's forgotten family. *Nature* 534: 151.
Nel A, Roques P, Nel P, Prokin AA, et al. October 2013. The earliest known holometabolous insects. *Nature* 503: 257–261. doi: 10.1038/nature12629.Epub.
Renfrew C. 2006. Becoming human: The archaeological challenge. *Proc Br Acad* 139: 217–238.
Renfrew C. 2008. Neuroscience, evolution and the sapient paradox: The factuality of value and of the sacred. *Philos Trans R Soc B* 363: 2041–2047. doi:10.1098/rstb.2008.0010.
Renfrew C, Frith CD, Malafouris L. 2008. Introduction. The sapient mind: Archaeology meets neuroscience. *Philos Trans R Soc B Biol Sci* 363:1935–1938.
Romanello M, di Napoli C, Gren C, Kennard H, et al. 2023. The 2023 report of the Lancet countdown on health and climate change: The imperative for a health-centred response in a world facing irreversible harms. *Lancet.* doi: https://doi.org/10.1016/S0140-6736(23)01859-7.
Shen H. 2013. Precision gene edition paves way for transgenic monkeys. *Nature News.*
Smith CM. January 2013. Starship humanity. *Sci Am* 308 (1).
Tattersall I. 2004. The dual origin of modern humanity. *Coll Antropol* 28, Suppl 2: 1–8.
Tattersall I. 2010. Human evolution and cognition. *Theor Biosci* 129: 193–201. doi: 10.1007/s12064-010-0093-9.
Tattersall, I. 2017. How can we detect when language emerged? *Psychon Bull Rev* 24: 64–67. https://doi.org/10.3758/s13423-016-1075-9.

# 10

# LONG-TERM SOCIAL IMPACT OF DOMINANCE PRIORITIES

> *A world where nearly one out every ten people lives in abject poverty is not a just world. It dooms generations to an existence that no one should be willing to tolerate.*
>
> Indermit Gill, Chief Economist, World Bank Group (Correcting Course, Poverty and Share Prosperity 2023)[1]

Dominance exerted through political, economic, social class, or military power adds privileged structures to social construction. Each sets up their own priorities to maintain a privileged, dominant position. This profoundly impacts several aspects of social dynamics (e.g., access to education, poverty rates) and individual growth opportunities, some with grave long-term impacts on personal development and equal rights issues. The following considerations address the impact on social aspects, early education, and problems related to equal rights and social progress.

The World Bank (2018),[2] applies the concept of the *multidimensional poverty measure*, which seeks to understand poverty beyond monetary deprivations and includes access to education and basic infrastructure along with the monetary headcount ratio at the $2.15 international poverty line of 2017 purchasing power parity, as one of the dimensions. According to the said report,

> Globally, inadequate access to sanitation stands out as the most common deprivation, affecting nearly a quarter (23.2 percent) of the population … Educational attainment (12.6 percent) and access to electricity (12.1 percent) are the most common deprivations globally after sanitation.

This report also emphasizes that 52 per cent of the population in sub-Saharan Africa experiences multidimensional poverty. It adds that the *global multidimensional poverty* headcount ratio for circa 2018 was 14.5 per cent, which exceeds the *monetary poverty* measure of 8.8 per cent.

DOI: 10.4324/9781003571544-11

The United Nations has addressed this issue through the report *Sustainable Development Goals* (2023),[3] which states that approximately 10 per cent of the world's population still lives in extreme poverty. Most survive with less than 1.90 US dollars a day and live in sub-Saharan Africa.

The *Poverty and Shared Prosperity 2022: Correcting Course* report from the World Bank Group stated that in 2020, the number of people living below the extreme poverty line rose by over 70 million. It further adds that over 3 billion people live on less than US$ 6.85 per day, which is the average of the national poverty lines of upper-middle-income countries. Under this view, poverty persists well beyond Africa. This report projects that 7 per cent of the world's population—roughly 574 million people—will still live in extreme poverty in the year 2030.

According to Serrano (*The Economist*, 2021), 85 per cent of the world's population lives on less than 30 US dollars daily.[4]

The report *Global Progress in Reducing Extreme Poverty Grinds to a Halt* provides a further reference on future perspectives from the World Bank (2022). It states[5] that an estimated 719 million people subsisted on less than $2.15 a day by the end of 2020, and predicts that by the year 2030, approximately 600 million people will survive on less than $2.15 a day. The report focuses on sub-Saharan Africa, which accounts for 60 per cent of all people in extreme poverty, estimated at 389 million.

A particular developmental age, i.e., early infancy, represents a sensitive period for food poverty. In its 2024 *Child Nutrition Report (Child Food Poverty)*, the United Nations Children's Fund (UNICEF) reported that, besides unhealthy foods becoming part of their diets, 'Globally, one in four children (27 percent) are living in severe child food poverty in early childhood, amounting to 181 million children under 5 years of age.'

The above briefings depict a sombre future for a significant proportion of the world population, concentrated in certain areas. Another consideration should be to stress the cumulative effect of poverty rates on human developmental opportunities (for example, Hanson (2023) on US dropouts). The single-year figures should be added for several previous years to build a composite—cumulated—of the complex dimension of poverty and its impact on individual and group cognitive development and access to individual self-opportunities considering the demands of the modern world.

The above considerations lead to an unavoidable consequence for educational development. As a point on this dimension, according to UNESCO (2023):[6]

> There are about 263 million children out of school. This includes 61 million children of primary school age (5 to 11), 60 millions of lower secondary school age (11 to through 14), and the first ever estimate of upper secondary school age at 142 million (15 to 19).

At the national level, the highest percentages of children out of school are reported from Somalia, South Sudan, Niger, Burkina Faso, Mali, Afghanistan, Syria, and several further African countries. According to the 2023 report from the *UNESCO Institute for Statistics* (*Data for the sustainable development goals*), in most sub-Saharan African countries with available data, fewer than half of adults have attained primary education,[7] which contrasts with over 75 per cent of the population achieving a primary education in most countries in the rest of the world, including Northern Africa.

Though significant deficits are seen in African countries, there is a significant spectrum of cultural demands per nation, e.g., the incidence of school dropouts in the USA (in 2023) involved 32.9 per cent of undergraduates who did not complete their degree programme. Additionally, 40 million Americans were college dropouts in a July 2021 report from the Borgen Project.[8] Thus, the data indicates that the educational problem attains differing social dimensions according to national socioeconomic profile standards and cultural demands. This also takes account of a direct correlation between lower income and increased dropout risk among all post-secondary students, as referred to in the US data. In the USA, according to the year 2023 report,[9] considering ethnic pertinence:

> Asian students have a 2.6% dropout rate. Black students have a dropout rate of 14.4%. White students are the most likely to drop out, at 33.4% compared to all other college dropouts.

The problem of early school dropouts attains a world scale, as shown by 2022 data from the *UNESCO Institute for Statistics*[10] reported in developing nations, where more than 64 million primary school students dropped out of their education in 2020, with significant differences among low- and middle-income countries.

In 2022, 11.1 per cent of young men and 8.0 per cent of young women in the European Union were early leavers from education and training. The proportion of early leavers from education and training in the European Union in 2022 ranged from 2.3 per cent in Croatia to 15.6 per cent in Romania.[11]

In some countries, one significant reason for early educational dropout relates to child labour: according to UNICEF estimates, 1 in 10 of all children worldwide is a victim of child labour.[12]

The previous comments lead us back to developmentally early socioeconomic and cultural conditions that contribute to setting up differential neurocognitive developmental profiles, as discussed in previous chapters. It seems reasonable to stress some socioeconomic and cultural vectors that condition neurocognitive development.

Given the fragility and plasticity of the infant brain and its defencelessness, the risks for the young remain in force but are now subject to additional variables generated by each individual. This is due to differential social parenting, a behaviour modulated by sociocultural dynamics and public policies that often results in marginalization and poverty, with consequences for health and cognitive development. These conditions carry the risk of brain and mental damage of social origin (Colombo, 2010). A risk that, if we approach it with an evolutionary criterion, represents a counter-revolutionary immorality to the extent that it violates one of the basic premises that allowed the survival of the human species. That is, a child's upbringing following the circumstances posed by the full development of a baby with high survival demands and unable to achieve self-sufficiency for a long postnatal period, a concept involving social transitions in human evolution (Foley and Lee, 1991; Foley and Gamble, 2009). Hence, the emergence of *Homo sapiens* did not occur as a necessary and culminating construction of natural history but as a product of a dynamic, interactive radiation that recognizes numerous demands adapted to different ecological niches.

Herrman and Call (2007) stated that phylogenetically close species do not share human cognitive abilities, involving a 'sociocultural intelligence' that tends to solve efficiently more complex mental operations. This implies a genetic domain in addition to the crucial prolonged postnatal period of brain organization during which complex social skills and instruments are acquired, allowing for rich social interaction. Such brain/mental development implies a significant energy consumption that amounts to 80 to 50 per cent of the total body requirement during the neonatal period and early childhood and 25 per cent on average during adult life (Leonard and Robertson, 1992, 1994; Leonard et al., 2003). The mentioned energy requirements place an initial high metabolic demand that should be met to obtain optimal cognitive development based on the demands posed by language and the learning of specific practices linked to particular human cultural conditions of an 'ultra-social' animal.

Given the energy requirements mentioned above, the percentage of body reserves—as body fat—is essential during early postnatal development and the transition from breastfeeding to the intake of solids. Under normal conditions, there is a significant load of reserves during childhood, although comparatively, at this stage show a lower muscle mass load (Leonard and Robertson, 1992; Leonard et al., 2003). Should those reserves not be met, there is a profound risk to attaining normal physical and cognitive development posed by low birth weight, early malnutrition, and early and inadequate replacement of breastfeeding. In this regard, on a comparative basis, according to Leonard and Robertson (1992) and Leonard et al. (2003), humans have the highest proportion of body fat during the postnatal period. This has been interpreted as an adaptation of the species by providing an energy reserve for the enormous and prolonged initial requirement of the human brain, combined with that infant's extended period of dependency.

The prolonged postnatal immaturity of *Homo sapiens* would not have allowed its survival as a species without belonging to a community organized to contain, educate, and provide extended parental care to the offspring. This extended period of postnatal development and immaturity allows for a prolonged phase of social/cultural moulding of the brain and mind, which involves neurobiological aspects linked to the formation of nervous circuits and of preferential neo- and palaeocortical 'synaptic pathways' related to cognitive and affective behavioural domains.

This postnatal slow cultural/social moulding of the brain and mental activity requires special care for the conditions of its development until adolescence. It is already unjustifiable to ignore the scientific evidence that points out the impact of upbringing conditions on brain development—at the level of its microstructure, neurochemistry, and dynamic organization—and cognitive performance, as referred to in previous pages. Without the body energy reserve (body fat), both the risks of supply during the transition from breastfeeding to the intake of solids means, as well as during exposure to any incidental demand, deprivation, or intercurrent illness, would increase the likelihood of interfering with the provision of the enormous calories required by brain development. Hence, there is a profound risk posed by low birth weight, early malnutrition, and early and inadequate replacements of breastfeeding, besides the immunological and emotional benefits that breastfeeding has.

Thus, one of the grave consequences of the deepening of social marginalization and the impoverishment of human conglomerates is its impact on cognitive performance from an early age. Indeed, the upbringing environment significantly affects subsequent child cognitive performance, which implies the compromise of the brain and functional circuits of *executive behaviour*. Hence, the conditions of poverty or underfeeding during the infancy period imposed on communities of the world's population constitute an ideological and ethical problem. Beyond its individual and social impact, it provides profit grounds to clientelist political strategies and objectives of maintaining an idle, unskilled workforce that lowers the cost of production. In such a regard, the abysmal distance that continues to be generated among human populations is additionally related to the acceleration of technological development and specialized labour competitiveness. This leads to a surplus of labour and a progressive decrease in its value.

Under the above considerations, the statistical data mentioned earlier on worldwide poverty rates accumulated over several decades have a crucial impact on the actual and future conditions of individual and social development of several communities.

## Notes

1 https://www.worldbank.org/en/news/press-release/2023/06/06/global-economyon-precarious-footing-amid-high-interest-rate.
2 https://www.worldbank.org/en/topic/poverty/brief/multidimensional-poverty-measure.
3 https://www.un.org/sustainabledevelopment/es/.
4 https://www.eleconomista.es/economia/noticias/11087781/03/21/El-85-de-la-poblacion mundial-vive-con-menos-de-30-dolares-al-dia.html.
5 https://www.worldbank.org/en/news/press-release/2022/10/05/global-progress-in-reducingextreme-poverty-grinds-to-a-halt.
6 https://worldtop20.org/out-of-school-children/.
7 https://uis.unesco.org/en/home.
8 https://borgenproject.org/2021-report/.
9 https://educationdata.org/college-dropout-rates.
10 https://borgenproject.org/early-school-dropouts/.
11 https://ec.europa.eu/eurostat/statisticsexplained/index.php?title=Early_leavers_from_education_and_training.
12 https://data.unicef.org/resources/child-labour-2020globalestimatestrends-and-the-road-forward/.

## References

Colombo JA. 2010. *Somos la especie equivocada?* (*Are humans the mistaken species?*). Eudeba. https://tinyurl.com/lsdozicg.

Foley RA, Gamble C. 2009. The ecology of social transitions in human evolution. *Philos Trans R Soc Lond B Biol Sci* 364: 3267–3279. doi: 10.1098/rstb.2009.0136.

Foley RA, Lee PC. 1991. Ecology and energetics of encephalization in hominid evolution. *Philos Trans R Soc B* 334: 223–231. https://doi.org/10.1098/rstb.1991.0111.

Hanson M. 2023. *College dropout rates.* Initiative.

Herrmann E, Call J. 2007. Humans have evolved specialized skills of social cognition: The cultural intelligence hypothesis. *Science* 317: 1360–1366. doi: 10.1126/science.1146282.

Leonard WR, Robertson ML. 1992. Nutritional requirements and human evolution: A bioenergetics model. *Am J Hum Biol* 4: 179–195. doi: 10.1002/ajhb.1310040204.

Leonard WR, Robertson ML. 1994. Evolutionary perspectives on human nutrition: The influence of brain and body size on diet and metabolism. *Am J Hum Biol* 6: 77–88. https://onlinelibrary.wiley.com/doi/abs/10.1002/ajhb.1310060111 1/2.

Leonard WR, Robertson ML, Snodgrass JJ, Kuzawa W. 2003. Metabolic correlates of hominid brain evolution. *Comp Biochem Physiol A Mol Integr Physiol* 136: 5–15. doi: 10.1016/s1095-6433(03)00132-6.

United Nations Children's Fund (UNICEF). June 6, 2024. *Child Food Poverty: Nutrition deprivation in early childhood*. UNICEF.

# 11

# BRIEF ACCOUNTS OF DOMINANCE EPISODES ACROSS HISTORY

As discussed before (Colombo, 2022, 2024a, 2024b) and in previous chapters in this volume, one basic behavioural profile of living organisms is based on dominance, as it conditions individual and species survival involving feeding, reproduction, physical fitness, and social standing. This premise implies competition and/or strategic alliances aimed at survival, whether physical, hierarchical, or—at the human level—theoretical or conceptual. Among possible options, resolution has—more often than not—involved terminal physical encounters, involving, at the unicellular level, the processes of phagocytosis, death of the opponent at the level of organized multicellular organisms (as a result of the interaction as prey, predator, or scavenger), and, at organized social levels, loss of social hierarchy, or subordination, or multiple forms of repression, as documented in anthropological and social studies (discussed and illustrated in Colombo, 2024a).

Though most of these hostile events have not reached historical records yet, a proposed listing (shown in Box 11.2, adapted from Colombo, 2024a) of recorded combat events reflects the perdurance of ancient drives and search for dominance.

**BOX 11.1 LIST OF WARS BY ERA[1]**

- Prehistoric warfare
- Ancient warfare
  - Ancient Greek warfare
  - Aztec warfare
  - Celtic warfare
  - Dacian warfare
  - Endemic warfare
  - Gaelic warfare
  - Gothic and Vandal warfare
  - Illyrian warfare
  - Mayan warfare
  - Roman warfare
  - Thracian warfare
- Medieval warfare
  - Anglo-Saxon warfare
- Early modern warfare
  - Napoleonic warfare
- Industrial warfare
- Modern warfare

DOI: 10.4324/9781003571544-12

**BOX 11.2 LIST OF WARS ORDERED CHRONOLOGICALLY BY THE YEAR THAT HOSTILITIES WERE INITIATED ACCORDING TO THE EDITORS OF *ENCYCLOPAEDIA BRITANNICA*. FURTHER EVENTS, LISTED IN ITALICS, HAVE BEEN ADDED BY JA COLOMBO (AS MODIFIED FROM COLOMBO 2024A).**

1300–1200 BCE
- Trojan War (dates uncertain)

1200–1100 BCE
- Trojan War (dates uncertain)

800–700 BCE
- First Messenian War (c. 735–715 BCE)
- Lelantine War (c. 720–680 BCE; dates uncertain)

700–600 BCE
- Lelantine War (*c.*720–680 BCE; dates uncertain)
- Second Messenian War (c. 660 BCE)

500–400 BCE
- Greco-Persian Wars (492–449 BCE)
- Peloponnesian War (431–404 BCE)

400–300 BCE
- Lamian War (323–322 BCE)

300–200 BCE
- First Punic War (264–241 BCE)
- Second Punic War (218–201 BCE)

200–100 BCE
- Third Punic War (149–146 BCE)

100 BCE–100 CE
- Gallic Wars (58–50 BCE)

600–700 CE
- Jinshin-no-ran (672)

1000–1300 CE
- Norman Conquest (1066)
- Crusades (1095–1291; sporadically thereafter)
- Gempei War (1180–1185)
- Barons' War (1264–1267)

1300–1400 CE
- Hundred Years' War (c. 1337–1453)
- War of the Eight Saints (1375–1378)

1400–1500 CE
- Hundred Years' War (c. 1337–1453)
- Thirteen Years' War (1454–1466)
- Wars of the Roses (1455–1485)
- Ōnin War (1467–1477)

1500–1600 CE
- Count's War (1534–1536)
- Araucanian Wars (1541–1558)
- Livonian War (1558–1583)
- Eighty Years' War (1568–1648)
- War of the Three Henrys (1587–1589)

1600–1700 CE
- Eighty Years' War (1568–1648)
- Kalmar War (1611–1613)
- Thirty Years' War (1618–1648)
- Powhatan War (1622–1644)
- Bishops' Wars (1639; 1640)
- English Civil Wars (1642–1651)
- First Northern War (1655–1660)
- War of Devolution (1667–1668)
- King Philip's War (1675–1676)
- War of the Grand Alliance (1689–1697)
- King William's War (1689–1697)

1700–1800 CE
- Second Northern War (1700–1721)
- War of the Spanish Succession (1701–1714)
- War of the Emboabas (1708–1709)
- Carnatic Wars (1746–1748; 1749–1754; 1758–1763)
- Queen Anne's War (1702–1713)
- Yamasee War (1715–1716)
- War of the Polish Succession (1733–1738)
- War of Jenkins' Ear (1739–1748)
- War of the Austrian Succession (1740–1748)
- King George's War (1744–1748)
- French and Indian War (1754–1763)

- Silesian Wars (1740–1742; 1744–1745; 1756–1762)
- Seven Years' War (1756–1763)
- Lord Dunmore's War (1774)
- Rohilla War (1774)
- American Revolution (1775–1783)
- First Maratha War (1775–1782)
- War of the Bavarian Succession (1778–1779)
- Cape Frontier Wars (1779–1879)
- French Revolution (1787–1799)
- French revolutionary wars (1792–1801)

1800–1900 CE

- Cape Frontier Wars (1779–1879)
- French revolutionary wars (1792–1801)
- War of the Oranges (1801)
- Tripolitan War (1801–1805)
- Second Maratha War (1803–1805)
- Third Maratha War (1817–1818)
- Napoleonic Wars (1803–1815)
- Black War (1804–1830)
- Peninsular War (1808–1814)
- War of 1812 (1812–1815)
- Creek War (1813–1814)
- War of Greek Independence (1821–1832)
- Padri War (1821–1837)
- Naning War (1831–1832)
- Pastry War (1838–1839)
- Mexican American War (1846–1848)
- Crimean War (1853–1856)
- Bleeding Kansas (1854–1859)
- American Civil War (1861–1865)
- War of the Triple Alliance (1864/1865–1870)
- Seven Weeks' War (1866)
- Selangor Civil War (1867–1873)
- Franco-German War (1870–1871)
- Acehnese War (1873–1904)
- Red River Indian War (1874–1875)
- Serbo-Turkish War (1876–1878)
- Anglo-Zulu War (1879)
- War of the Pacific (1879–1883)
- Gun War (1880–1881)
- Sino-French War (1883–1885)
- Serbo-Bulgarian War (1885–1886)
- Sino-Japanese War (1894–1895)
- Spanish-American War (1898)
- Philippine-American War (1899–1902)
- South African War (1899–1902)
- The War of a Thousand Days (1899–1903)

1900–2000 CE

- Acehnese War (1873–1904)
- Philippine-American War (1899–1902)
- South African War (1899–1902)
- The War of a Thousand Days (1899–1903)
- Boxer Rebellion (1900–1901)
- Moro Wars (1901–1913)
- Russo-Japanese War (1904–1905)
- Pig War (1906–1909)
- Mexican Revolution (1910–1920)
- Italo-Turkish War (1911–1912)
- World War I (1914–1918)
- Baltic War of Liberation (1918–1920)
- Russian Civil War (1918–1920)
- Russo-Polish War (1919–1920)
- Rif War (1921–1926)
- Chaco War (1932–1935)
- Italo-Ethiopian War (1935–1936)
- Spanish Civil War (1936–1939)
- Sino-Japanese War (1937–1945)
- Phoney War (1939–1940; no actual hostilities)
- Russo-Finnish War (1939–1940)
- World War II (1939–1945)
- Greek Civil War (1944–1945; 1946–1949)
- Arab-Israeli wars (1948–1949; 1956; 1967; 1973; 1982)
- Korean War (1950–1953)

- Algerian War (1954–1962)
- Vietnam War (1954–1975)
- Six-Day War (1967)
- War of Attrition (1969–1970)
- Yom Kippur War (1973)
- Dirty War (1976–1983)
- Afghan War (1978–1992)
- Iran-Iraq War (1980–1988)
- Falkland Islands War (1982)
- Persian Gulf War (1990–1991)
- Bosnian conflict (1992–1995)
- Kosovo conflict (1998–1999)
- *Second Congo War (1998–2003)*

2000– …CE

- Afghanistan War (2001–2014)
- Iraq War (2003–*2011*)
- *Iraq Civil War (2013–2017)*
- Syrian Civil War (2011–*2022*)
- *Yemeni Civil War (2014–)*
- *Russian-Ukrainian War (2014–)*
- *Israeli-Middle East War (2023–)*

The list in Box 11.3 updates additional hostilities, not including—except for Mexico and Colombia—the North, Central, and South American continents.[2]

**BOX 11.3. CURRENT WARS AND CONFLICTS (2023).**

| | | |
|---|---|---|
| Afghanistan | Ghana | Somalia |
| Algeria | Iraq | South Sudan |
| Benin | Ivory Coast | Sudan |
| Burkina Faso | Mali | Syria |
| Cameroon | Mauritania | Tanzania |
| Central African Republic | Mexico | Togo |
| Chad | Mozambique | Tunisia |
| Colombia | Myanmar | Uganda |
| Democratic Republic of the Congo | Niger | Ukraine |
| | Nigeria | Yemen |
| Ethiopia | Russia | |

Killing for feeding, territorial, or social prevalence purposes has characterized natural evolution, whether associated with alliances or not. At the human level, these universal, basic drives acquired a different dimension and moral consideration in terms of humane survival conditions, combined with social constructs aimed at expressing solidarity—also represented by other animal species—and/or physical and social gains and aesthetic values that introduced a cultural/cognitive level. Additional listings[3] by the *Geneva Academy*, *Relief Web*, and *Our World in Data* tend to complete the world picture of armed conflict, characterized by periodic regional warfare and an increasing level of massive killing weapons.

The chart shown in Figure 11.1 reflects the status of nuclear-armed capabilities.

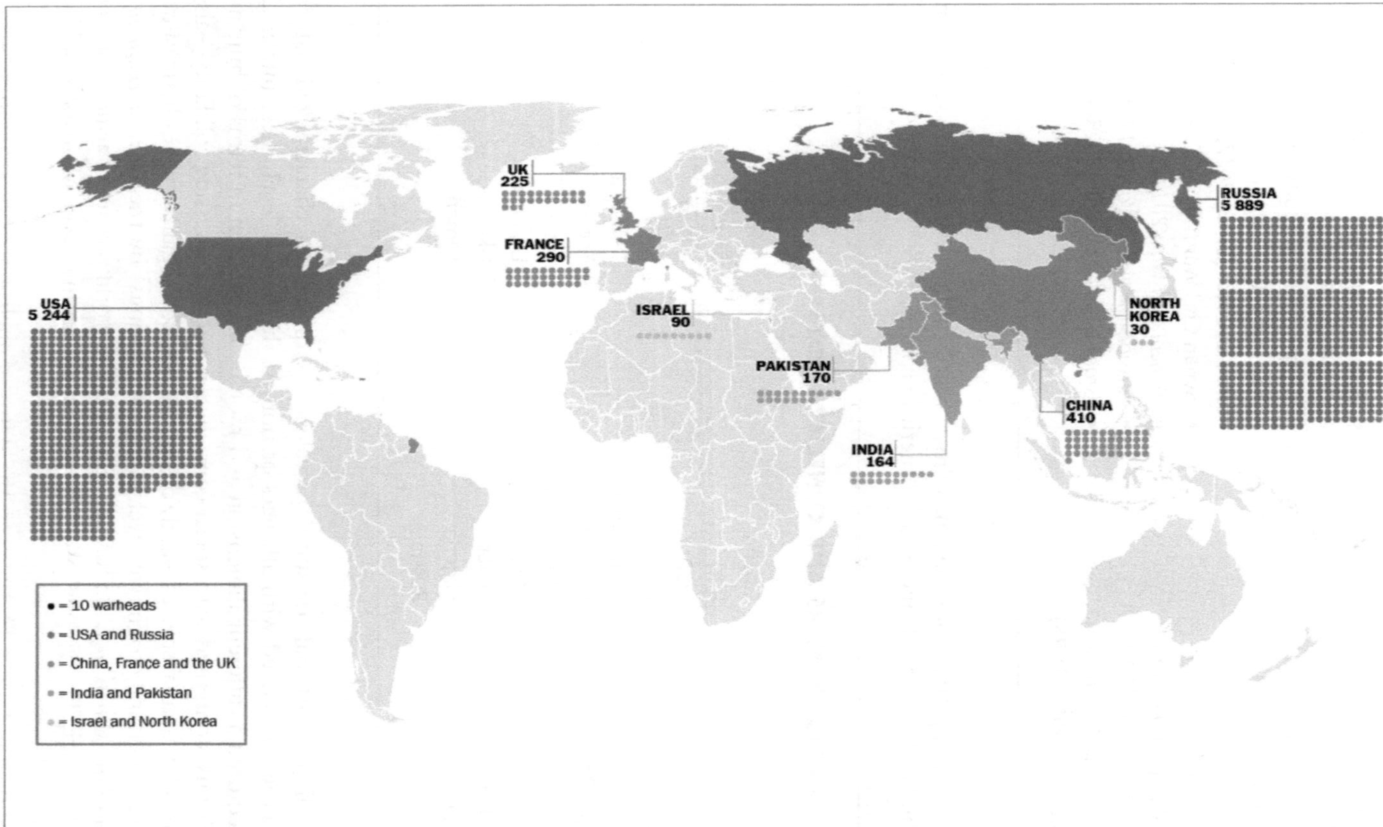


**FIGURE 11.1** Global nuclear weapon inventories, January 2023.

*The boundaries used in this map do not imply any endorsement or acceptance by Stockholm International Peace Research Institute.*

*Stockholm International Peace Research Institute (SIPRI). Yearbook 2023.*

According to the same Report by SIPRI:[4]

> The nuclear arsenals of the other nuclear-armed states are even smaller, but all are either developing or deploying new weapon systems or have announced their intention to do so.
>
> The availability of reliable information on the status of the nuclear arsenals and capabilities of the nuclear-armed states varies considerably.

Though manifest reasons for generating or entering a conflict are the expression of historical-cultural contexts and political strategies, the fundamental behaviour of dominance or strategic gain is anchored in ancient drives for survival and predominance, whether of territorial, feeding resources, physical dominance, or ideological nature. This bonding to ancient allegiances at the individual level is anchored at the construction of identity contexts closely bound to circumstantial convenience or to parental psychological anchoring, and then transferred to a collective ego.

Despite public humanitarian declarations by national leaders and international organizations, current human development is taking a toll on human lives and future growth, including child death and famine, because of the consequences of current wars (see Box 11.3), in addition to ethnic conflicts resulting in civilian deaths:[5]

> The military conflict in Sudan is fuelling 'a humanitarian emergency of epic proportions' that threatens the entire country, the UN humanitarian chief warned on Friday.
>
> … the African country 'could tip the entire region into a humanitarian catastrophe.' … hundreds of thousands of Sudanese children are severely malnourished and at imminent risk of death if left untreated.

In Sudan, as an example of ongoing conditions affecting life caused by ongoing wars, food insecurity and school disruption add to the harmful social impact of current events. In such regard, projections based on Johns Hopkins' Lives Saved Tool modelling would indicate that at least 10,000 children aged under five years may have died by the end of 2023 due to an increase in food insecurity and disruption to essential services, including schooling, where about 19 million children were awaiting the reopening of schools.

These consequences are reproduced in other countries due to war or aberrant social conditions, which reflects the reluctance of dominant political powers to reconsider the conditions of their supremacy.

## Notes

1 https://en.wikipedia.org/wiki/Prehistoric_warfare.
2 According to the map found at https://en.wikipedia.org/wiki/Prehistoric_warfare.
3 https://geneva-academy.ch/galleries/today-s-armed-conflicts; https://reliefweb.int/report/world/conflict-trends-global-overview-1946-2022; https://ourworldindata.org/grapher/ongoing-wars.

4 https://www.sipri.org/yearbook/2023/07.
5 https://allafrica.com/stories/202311060124.html.

## References

Colombo JA. 2022. *Dominance behavior: An evolutive and comparative perspective.* Springer Nature.
Colombo JA. 2024a. *Evolution and the human-animal drive to conflict. A psychobiological perspective.* Routledge.
Colombo JA. 2024b. *Human biome and human behaviour.* Routledge.

# 12

# MENACES TO HUMAN CREATIVENESS

## Creativeness Should Be Considered a Social Value

The previous analysis has an additional inbred consequence: a menace towards fostering one main characteristic of the human species—its potential creativeness in every life dimension. It cannot be reviewed without considering *ab initio* human construction of the concept of 'freedom' or, more accurately, human 'degrees of freedom' as discussed in Colombo (2013). Most critical is self-awareness of the construction of inhibitory codes and rules in human *socialization*, which this author finds too close to *domestication* and fencing creativity. Social construction implies specific rules on human interaction and convivence, though subject to the authoritarian expression of political and economic drives. Hence, besides probabilities of access to adequate educational levels and the construction of individual self-social awareness, authoritarian social structures constrain the implied restriction of personal freedom.

Each species has survival mechanisms adapted to their structural and physiological needs and natural niche in the natural kingdom. In all of them, without exception, some conditions reduce their degrees of freedom according to the characteristics of their species and the expression of a behavioural phenotype that tends to meet their reproductive and survival needs. An unstable internal conflict/tension ensues whenever it does not satisfy the demands. This provides the basis for the ever-changing evolutive conditions and species adaptations. In human terms, when we deal with the concept of 'freedom', we are inclined to ignore the limitations based on the heritage from our ancestral origin, biological construction, and personal sociocultural history. Thus, based on operative terms, biological and social variables condition our range of true freedom. Hence, the crucial point would reside on the social contexts that could maximize freedom of expression; thus, unless these conditions are considered, we would fall on a somewhat symbolic construction of the meaning of freedom, a literary utopic construction.

Within individual '*freedom*', an epistemological trap evolved from the traditional dichotomy between brain and mind and body and *spirit*. This divergence generated implicit

DOI: 10.4324/9781003571544-13

dualism and *a priori* rejection of integrated arguments that were considered reductionist in a derogatory manner. To some extent, this emerged due to the lack of conceptual bridges or theoretical models of the brain that would link behavioural domains with its complex molecular and electrochemical organization and neural, conditional micro-connectivity. Further analytic considerations on the anthropological field of this complex theme were made in Colombo (2020).

Hence, in real-life terms, how could enhanced conditions for freedom of thought potentiate human *creative* drives? Unfortunately, this quest for creative freedom conditions is not met in public policies nor does it represent a significant concern within a dominant section of our public human endeavours. Within the organic dimension, the characteristics of the *Homo sapiens* brain would not depend solely on the structural modification of neural connectivity of specific regions/cortical areas but also on the generation/reformulation—neuroplasticity—of distributed neural circuits. In the social dimension, besides traditional learning conditions, poverty and marginality contribute to reducing the phenotypic richness of the species, enhancing the trend towards dominance by unique and prevalent thinking formats and, therefore, representing counter-evolutionary factors. In other words, it derives into cognitive deprivation on the construction of identity, the access to critical knowledge, and the construction and expression of the *creative* potential (Colombo, 2007, 2015).

The above-described condition involves deep social imbalances in cognitive development and the construction of identity, access to critical knowledge, and the construction and expression of creative potential. Consequently, it represents a counterargument against the concept of a *globalized* world—contrasting with the increased access to information by selected population strata. In human evolutive terms, plastic, imaginative, and productive behaviour, captured in our cultural heritage, increased the number of possible survival alternatives and further divergent evolutive perspectives. Phenotypic, ethnic, and cultural varieties are intrinsic components of our adaptive capacity as a species and a potential source for the exercise of dominant bouts among populations and cultural expressions. Regarding the cultural aspects of the *Homo* evolution, one must bear in mind the possible hiatus or discontinuities that stem from the fast degradation/replacement processes of brain connectivity in addition to the methodological limitations when attempting to establish correlations between brain development and the emergence of cultural objects in ancient human samples.

> Children living in poverty generally perform poorly in school, with markedly lower standardized test scores and lower educational attainment. The longer children live in poverty, the greater their academic deficits. These patterns persist to adulthood, contributing to lifetime-reduced occupational attainment … The influence of poverty on children's learning and achievement is mediated by structural brain development.
>
> *(Hair et al., 2015)*

As mentioned in previous chapters, human development is under menacing or distorting conditions. They are represented by social inequities, dominant sociopolitical bouts, reiterative war conditions, and a predominant economic goal of potentiating the supposed need for consuming commodities that benefit economic market dominance. In this regard, the development of financial and sociopolitical structures has led *Homo sapiens* to develop a segmented world community. The consumer drive is fuelled by current public pressure imposed by oligopoly and monopoly interests that manage diverse publicity tactics to promote and incentivize consumer behaviour at the expense of health security and access to critical knowledge for buying decisions, with increased ecological risk in consumer goods production.

In modern times, within this consumerist social atmosphere, a central problem in any social construction—especially at all levels of formal education—is how to promote and express creativity, since it usually carries a potentially disruptive or confrontational component. Thus, it represents a significant social and educational challenge.

The above considerations imply a conflictive alternative regarding how to deal pedagogically with the promotion of inventiveness and creativity when they represent an actual threat to current dominating thoughts and the massive consumer process offered by oligopolies and monopolies. Promoting imagination and creativity implies a state of mind and an attitude towards learning and knowledge that does not easily reconcile with the requirements of a consumer society, conditioned by cultural profiles of social structures aimed at massive consumption of goods and information.

The previous statements conform to a social condition moulded towards a consumerist behaviour executed under a fog of calculated information that tends to cancel critical decision processes and reduce creative and argumentative (dissenting) behaviours.

Under this perspective, the crackled image of the world population and cultural traditions and beliefs do not suggest a uniform process and level of information and social developmental degrees, nor equal cultural and economic stances, that would suggest the promoted concept of *globalization*. Even further, religious and cultural traditions show periodic rounds of clashes among traditionalists, capitalists, and imperialists between and within intercultural bands, with local wars and the menace of more global ones sparkling at each turnover of the calendar year. This crackled vision augurs an impact on people's lives and our menaced environmental and ecological universe.

The dominance profile embedded within living species has not abandoned our most enacted public human behaviour, and involved decision makers. As discussed previously, dominance drives are most probably crouched within the folds of our essential brain masterpiece and expanded to various domains of human life.

In essence, in human history dominance represents an unending series of battles and wars of increasing damage, cultural replacement, physical cruelty, economic suffocation, and condemnation that goes well beyond basic animal survival strategies—it is as if that ancient, primal survival-based animal behaviour evolved under a magnifying and distorting lens. While the multiple and variable animal species, including early *Homo* species, exert some limited territorial or ecologic dominance, the *sapiens* version has managed to put at stake world living conditions, potentially affecting the survival of all macro species and resetting the interaction of life forms and living conditions condensed in the concept of *Anthropocene*.

## References

Colombo JA. 2007. *Acerca del desarrollo cerebral infantil* (*On brain child development*). Paidos Editorial.

Colombo, JA. 2013. *Bajo libertad condicionada* (*Under conditioned freedom*). Imago Mundi.

Colombo JA. 2015. *Los Homo sabios. Globalizados o segmentados?* (*The wise Homo. Globalized or segmented?*). Buenos Aires Books.

Colombo JA. 2020. *Creativity: A profile for human species.* Cambridge Scholars.

Hair, NL, Hanson, JL, Wolfe, BL, Pollak, SD. 2015. Association of Child Poverty, Brain Development, and Academic Achievement. *JAMA Pediatr* 169: 822–829. doi:10.1001/jamapediatrics.

# 13

# HUMAN BIPOLAR DRIVES

## Creativity vs. Dominance

> You, your joys and your sorrows, your memories and your ambitions, your sense of personal identity and free will, are in fact no more than the behavior of a vast assembly of nerve cell.
>
> (Crick, 1995)

> It is interesting to include the disturbing proposal that the process of human civilization involves one of self-induced human domestication.
>
> (Colombo JA, 2010)

Biological organisms share several common behaviours that impact the remaining species and ecological equilibrium. Except for catastrophic events of extra-biological origin, organisms depend on interactions with multiple biological orders and species and express a partial overlap of resources with different species with which they have positive or negative interactions. Throughout natural history, these have usually reached a natural equilibrium and have been limited to a given ecological niche, except when major environmental conditions impact the survival probability of several of them.

The emergence of a macro species with a comparatively progressive, creative capacity in cognitive domains has reset the conditions needed for such equilibrium. This set up a bipolar creature capable of the most appealing and creative performance yet able to express the most cruel and destructive behaviours. The survival of these drives is probably present in most humans, though not necessarily with identical weight or sharing the same caring and constructive drives. Under this hypothetical circumstance, the final behaviour would depend on emotional variables linked to individual histories of survival, education, convenience, or ideological profile. Thus, cultural impact expressed as either economic convenience or as ideological or religious dominant profiles will turn the balance in one of the following directions regarding solidarity, aggression, or indifference when subjected to life circumstances.

DOI: 10.4324/9781003571544-14

Under the general picture drawn, conditions that lead to severe inequalities, dominance, indigence, poverty, and social class priorities set up the degrees of individual and social integration or unrest. Under this scheme, dominance can obscure or obliterate the desirable expression of creativeness and solidarity and become a source of conflicts triggered by the decision of a comparative minority of economic and dominant interests. While this dominant social stratum plays its scheme of power profits, most workers and creatives keep working on the everyday basic conditions that attempt to make human life practically and emotionally worthwhile.

In this regard, the spread of sociocultural and informational areas allowed by a comparative minority integrated with economic and dominant interests only allows for a relative, tailored access to information by the public. Furthermore, it must be pointed out that this universe remains fractured and moulded by systematic propaganda and consumer choice manipulation, for the possibility of reaching their products does not necessarily acquire universal sociocultural access nor cognitive and emotional conditions to build a full conscience of the actual needs.

Thus, control of market propaganda and the relative costs of goods offered keeps fracturing the social web, which provides an additional source (for the further and progressive empowerment of minorities and segregation of individual opportunities) and access to a fully improved quality of life. Furthermore, such a fracture provides an environment that distorts the real individual capacity by limiting access to food and culture that maximizes brain/mental potential development and cognitive processing. Though the human universe offers a variety of brain/mental profiles, the result is that cognitive limitations are not independent of physical and sociocultural vectors but the consequence of a paucity of early adequate feeding and cultural stimuli.

It must be stressed that, since early times, individual physiological and behavioural differences have staged the conditions for individual aptitudes channelled towards dominance and the consequent building of stratified hierarchies. Yet, this represents a primal evolutive biological condition that human social minorities have potentiated through eons, promoting different sociocultural forms of dominance (Colombo JA, 2022) instead of building a solid perspective for individual cognitive growth, social expression, and collective gain in a multicultural environment. Incorporating intraspecific competition and combining cooperation and competition boosted technological development and defined political-financial strategies. Subdued by socioeconomic leaders, citizens remained bound to incorporate the interests and strategies of those leaders into their daily lives, potentiated by media and propaganda pressure. Consumer promotion generated conducts for which there was no previous training or experience. This sort of public *assault* by scientific-technological development, massive selling strategies and the prevalence of cultural idiosyncrasies led to a *technological-dependent* and consumerist culture that determined cultural paths of human development.

In such regard, maximizing material profit, as opposed to social recognition, has dominated human evolution, displacing the concept of limiting gain to the necessary life demands and social recognition. This material drive evolved to control public needs and conditioned behaviours to imprint a consumerist habit and supposed needs. As succinctly put forward by Eduardo Colombo (2013):

> If we abandon the utopian dimension of change, we will continue to uncritically repeat all the perversions of the system.

As stated earlier,

> Hence, the Gordian Knot that would need to be undone resides in cultural and socio-political idiosyncrasies and inertias, the so-called specific epistemic webs in which our societies have been construed. There are enough differences among them to anticipate that, ahead of us as species, there does not seem to be one single 'globalised' path, nor a single solution.
>
> *(Colombo JA, 2021)*

The basic ancestral, predatory animal social behaviour based on *dominance* or *prevalence* in non-human species is also expressed in the sophisticated human society, though utilizing a variety of political and cultural forms and styles that tend to disguise those basic ancestral drives. Though basic survival behavioural responses are based on *fight or flight*, these are expressed within the human cultural domains through social mannerisms and displaced behaviours. In addition to fight or flight, a further form of response resides in socialization and institutionalization, acceptance of rules that provide the means to buffer and modulate radical public responses, which lead to trails of reduced degrees of freedom. At the human level, this places the advantage of a multicultural humanity at risk since, beyond all consideration of solidarity, the suppression of cultural variety and the freedom to develop communities with alternative cultural offers would threaten behavioural variety, the basic condition for species' adaptive success and optimal survival. Therefore, suppression of cultural variety represents a counter-evolutive action since it tends to obliterate one of the primary characteristics for the survival of the human species: its ability to develop a rich menu of possible strategies. It is deeply immoral in that it puts the species' future at risk, a transcendent ethical value.

In their evolutionary journey, human communities built imaginary worlds and productive, emotional, and aesthetic structures that generated various operational proposals (behaviours) that have allowed them to survive in all corners of the planet since abandoning the African niche. This plastic, imaginative behaviour increased the number of possible alternatives for the species' survival. It should be considered that phenotypic and cultural variety is an inseparable component of the central engine of the adaptive capacity of *Homo sapiens.*

Human society ought to fully assume the demands derived from the human species' original biological basic construct and consider that human comparative biological advantage first lies in the extended period of postnatal brain immaturity. With it also comes a prolonged phase of brain development and social *moulding* of the mind.

However, despite those fundamental principles for survival—that is, those involved with early childcare—several modern, worldwide communities survive today in conditions that contradict each of these basic requirements by condemning their child populations to insufficient and inadequate conditions from the points of view of nutrition, health, and social containment. This potentially constitutes social damage to the childhood brain development, which will be a function of the depth of that condition, its persistence over time, and the recipient's age. The profound inequalities of child-rearing are the germ of a caste society, of a society without real freedom of social mobility.

> Educational public policies that advocate knowledge, the exercise of imagination and divergent thinking or conceptual expansion can only be effective if they are integrated into equity in social construction and the optimisation of individual freedom. Indigence, poverty, privilege and class prejudice are its powerful opponents.
>
> *(Colombo JA, 2007)*

> Human species has not only proved to be incapable of overcoming ancestral animal drives (combination of dominance/prevalence and unequal rights). Additionally, it applied its cognitive capacity to develop cultural and institutional means to cramp this animal drive to social structures also to impose it either financially, military, or by subtle or overt psychological media means. This action attempts to camouflage or obscure the true power domain and the generation of inequalities that deeply affect human development.
>
> *(Colombo JA, 2022)*

### Education as a Potential Generator of Host-Parasite-Like Induced Behaviour or Conditioned Behavioural Profiles: An Extended Concept

Let us metaphorically consider mental parasitism—any persistent, single concept or mental bias incorporated through the educational or social interactive process. This would be neurobiologically represented by a synaptic constellation able to persistently contribute to the construction or driving of thinking processes and behaviour, in the same way that, in nature, parasites drive insect behaviour. A couple of examples of the latter follow, as it has been recognised that in nature:

> Parasites modify a wide range of physiological, behavioural and morphological traits in their hosts.
>
> *(Poulin, 1999)*

More recently, Hughes and Kibersat (2019) stated:

> Some parasites have evolved the ability to precisely control the behaviour of animals in ways that enhance the transmission of parasite genes into the next generation. This is the concept of the 'extended phenotype' first conceived by Richard Dawkins in 1982. It states that the behaviour we observe in animals is due not only to the expression of their genes but also to the genes of parasites infecting them. In such cases, the behaviour is an extended phenotype of the parasite.

Additionally, Hafer-Hahmann (2019) stated:

> Ensuing benefits to the parasite can be accidental due to host responses or side effects, but the parasite might engage in true host manipulation, which alters its host phenotype, such as its behaviour, to increase its own fitness. Central to this definition is that the trait in question is controlled by the parasite to some degree.

Also, as reported by Golembiewski (2024):[1]

> Periodical cicadas lay their eggs in tree branches, and when those eggs hatch, the resulting baby cicadas, or nymphs, drop to the ground and burrow into the soil … At some point, the insects are exposed to spores of the fungus Massospora cicadina. The spores make their way into the cicadas' bodies, and from that point on, the cicadas are at the mercy of the fungal parasite.

These excerpts provide a metaphoric biological perspective of mental processes that could be affected by early education and/or social interactions. In this regard, two dimensions of non-conscious components should be considered at the human level. One derives from the neurobehavioral impact of early education and care from birth, added to conditions imposed by social and economic relative freedom. The other one relates to what has been generally accepted: the gut microbiome integrates a brain-gut microbiome interaction that affects behavioural drives (see review in Colombo JA 2024a, 2024b). Thus, human behaviour is not independent of interaction with commensal microorganisms, whose equilibrium disruption affects human behavioural processes and health conditions.

For eons, humans and their ancestors have contributed to ground bacterial conditions under primal living conditions through the deposition of microorganisms and modified undigested seeds, just as other species have. Thus, from an evolutive point of view, all species provide an intermediate stage in the recycling of soil biological compositions—involving a myriad of bacteria, parasites, and intermediate chemicals—which, in the case of humans, has been significantly suppressed under modern human sanitary conditions. On evolutionary grounds, it does not invalidate the concept that every biological organism, including humans, provides an intermediate stage in the organic renewal chain. In such a sense, from a biological point of view, humans may be included as an intermediate stage for maintaining the Earth's organic richness and renewal under natural conditions, if not because of its uncontrolled consumption and disruption of species balance. Thus, the human digestive system represents an incubator of many microorganisms whose balance depends on health conditions.

Thus, these biological examples could be extended to those potentially *parasitic units of cultures* characterized as *memes* by Richard Dawkins (1976) and considered parasites of the mind by Boudry and Hofhuis (2018). These parasitic constructions have contaminated the thought process since early human developmental days by affecting thought and emotional states that drive thinking patterns, as if a biological parasite would affect the behaviour of its host and other species in nature, as described above. These memes burden cultures, thus conditioning the creative process, also potentiated by cultural strategies as a form of subconscious cultural dominance.

It can be proposed that behavioural states and emotional processes that affect rational thinking involve the individual's complex biobehavioural interactions with resident *parasitic units of culture*, as parasitic organisms would act in nature. These interactions condition internal thoughts and open behaviour, and their disruption affects mental and emotional thoughts and behavioural processes. Through early educational and cultural instances, mental processes incorporate those basic behavioural patterns and interactive cultural units involved in behavioural construction. These mental processes are exposed to the iterative exposure of dominant political and commercial—manifest or subconscious—slogans that condition mass behavioural drives and opinion profiles.

Community dynamics acquired increasing complexity since farming entered the social horizon, contributing to population settlement and growth. The socialization of labour moulded a mind based on the complex interactions within and between groups. In it, the language provided an instrument for creating relational structures of various kinds (mystical, ideological, territorial), with the consequent further segmentation of the population universe. With the increase in the size of the communities, the notion of individual survival or equals progressively modified the community reference of each member. The community structure began to fragment according to different parameters. In general, and in metaphorical terms, *Homo sapiens faber*, *Homo sapiens nundinator* or *mercator* (*Homo sapiens economic* according to some authors), and *Homo sapiens pauper* increasingly occupied asymmetrical spaces of land, economic and military power, public income distribution, and developmental perspectives.

## Concerned Conclusions

> Seventy-five years after the United Nations' historic ratification of the Universal Declaration of Human Rights, the global commitment to protect the fundamental human rights of all people is under grave threat.
>
> (Beyrer et al., 2024)

To generate dominance or behaviour control, something to fear must be involved, if not rational adherence. This has been a recurring thought throughout civilization. It supported different forms, whether overt or covert, included in politics (latent or actual physical or financial repression), religion (punishing gods, esoteric menaces), education (forms of punishment and thought process conditioning), or any form of a relationship whose objective is to exert authority and control people's behaviour or social entropy. Its origin in its primary forms is ancestral since it manifests in the animal kingdom, taking the form of physical challenge, reproductive privileges, or social punishment.

Control or *status quo* results from a multiple-term equation, such as within social entities and international equilibrium, which depends on the relative strengths of multiple military or economic factors. Social control can be exerted through mass support and collective brainwashing, reinforced by technological dominance. The former has a local impact, while the latter sets up a macro stage that involves a complex strategic interactive factor. These bi-front operative event conditions usually result in a persistent instability of the local social system and international *status quo*.

As mentioned, behavioural dominance represents a primal, ancestral drive that, in modern times, is mediated by education, propaganda and a military power that conceals a high risk for non-belligerent countries. This includes the passive involvement of bystanders, often including countries conditioned by concealed interests or previous commercial treaties. As such, the world appears to be navigating troubled waters that override the multiple creative and constructive expressions that humanity can provide and use to justify its mere existence. Would subdued countries have the power to raise their voices enough to modulate international political unrest? So far, history does not give evidence in such regard.

The central problem behind this sombre perspective lies in the political and economic vested interests under the cover of high-sounding pleas for—barely responsive to—human rights and humanitarian demands. These seem to construct cynical arguments considering the worldwide statistics on human poverty and access to education and creative or productive life, as reported in previous chapters (and also in Colombo JA 2007, 2015, 2019, 2022, 2024a, 2024b). To generate a change in this self-destructive drive, a widespread claim to dislodge the representative puppetry controlling critical or nodal political decisions should be reached, representing a problematic goal under current conditions. Progressive, slow changes in the profile of political representatives represent a real risk of being co-opted by the same political tides.

Thus, human constructs of future *worlds* are loaded by similar basic behavioural drives that characterize animal history since its primal origin, involving behavioural prevalence and dominance. This places humans as dominated by a bi-front creature where dominance drives weigh more than their creative and solidary components. Hence, the survival of those components seems to be subjected to—and used by—the drivers of dominance.

The process of cultural 'homogenization'—functional to the consumerist objective—obtains strong allies in advertising campaigns and financial pressures, which maintain slogans of supposed needs and comparative welfare. In them, what is essential is confused with what is banal or superfluous. It is the mercantilist universe of the *Homo sapiens nundinator*. Excluded, the *Homo sapiens pauper* wanders between submission, criminal violence, and revolutionary drives.

If the final, absolute value of the principles of the prevalent ('fittest'?) communities is the survival of the species regardless of the values it supports or the survival conditions, then the concentration of power and the irrepressible scientific-technological progress—a powerful instrument of knowledge, turned into an instrument of domination—will continue its almost omnipotent development beyond any natural, social control or ethical consideration. Under those circumstances, differential access to educational levels keeps promoting the fragmentation of the human community towards a potentially irreversible change involving profound social stratification and differential access to increasing scientific and technological development rates.

Human societies have undergone critical changes in their structures and goals throughout history. However, those that would have seemed to have enclosed a potential for genuine democracy and mutual international dialectic coalescence so far have been or conditioned by dominant powers. These modulate people's opinions, drives, and alliances through the manipulation of access to goods in consumerist societies, misinformation through massive communication channels, caste dominance, state terrorism, and other forms that abort the emergence of actual sociopolitical changes.

Besides physical survival, humans must demand options for creative and productive conditions. Yet, current statistics show a slow correction of food and health variables that represent the bottom line for human beings and developmental conditions that significantly compromise the development of higher mental processes. Thus, the build-up of biological and cultural access to individual development in society has been at stake for decades. Its slow rate of change contrasts with the fast rate of technological advances. The gap grows wider for each generation that comes under those conditions, besides those that drift from optimal personal futures due to local cultural conditioning factors.

In such regard, the early-age developmental death toll is still marking a debt to the human future, even more so considering that surviving infants in several countries are deprived of any positive life prospects in terms of education, self-development, and productive social integration. As an example, in terms of child death rates, the following statement is from the 2023 UNICEF report:

> Each year, an estimated 4.9 million children around the world still die before their fifth birthday, and an additional 2.1 million children, adolescents and youth die at ages 5–24. These deaths are concentrated in sub-Saharan Africa and Southern Asia.[2]

As a side comment, one significant factor in attempting to analyse future employment and its social impact falls on artificial intelligence (AI). The emergence of AI applied to the automation of current occupations (major industry employment areas exposed to automation by AI in the Euro area would include clerical support workers, professionals, technicians and associated professionals, and managers) projects that, *globally*, 18 per cent of work could be automated by AI (Briggs & Kodnani, 2023[3]). According to the same report, two-thirds of current occupations could be partially automated by AI. Extrapolating our estimates globally suggests that generative AI could expose the equivalent of 300 million full-time jobs to automation.[4]

The emergence of this potent variable represents resetting social construction for several generations. This reinforces the growing requirement of labour access to training and the current worldwide disjointed access to education, which has had a significant impact on countries loaded with economically and educationally relegated populations for several generations, as described in previous chapters. Hence, the impact on the overall picture of socioeconomic rights brings up the ever-growing gap in educational access in a world driven by an increasing demand for specialized labour. This growing, uneven educational gap would reinforce the path of a disjointed socioeconomic world and its consequences on social inequality.

In the meantime, people are self-absorbed in surviving activities of different natures, whether physical, self-preserving, or creative. It seems that time is overdue for a call for political and cultural changes that could help the developmental future of a vast population majority while a privileged minority finds its way out through its financial, technological, and political resources.

Ancient animal survival drives persist in humans, masked under various behavioural paradigms. Fight and flight remain basic behavioural principles. Even subdued under religious or mystic beliefs, aggressive and defensive behaviours emerge to defend or fight for even the most sophisticated peaceful beliefs, and events throughout humankind's history support this evidence.

Moreover, the range of physical or emotional survivability now includes religious, material, and, in general, dominance forms in all orders of living matter, most dramatically on armed potential. In this regard, an example of the current level of potential nuclear arms capabilities is summarized in the following two excerpts:

> In 2023, the nine nuclear-armed states spent $10.8 billion (13.4%) more on their nuclear arsenals than the year before, a total of $91.4 billion.[5]
>
> Overall, the number of nuclear warheads in the world continues to decline. However, this is primarily due to the USA and Russia dismantling retired warheads. Global reductions of operational warheads appear to have stalled, and their numbers are rising again. At the same time, both the USA and Russia have extensive and expensive programmes under way to replace and modernize their nuclear warheads, their missile, aircraft and sub-marine delivery systems, and their nuclear weapon production facilities.[6]

One might wonder if humans have not contributed to distorting the concept that life is an open opportunity to one in which actions are nothing more than the expression that humans constitute an actual plague, capable of surviving—given enough time—under any circumstance and habitat, no matter how extreme, at the expense of the natural environment and their human partners or other species. In such respect, they opportunistically affect the rest of the species through actions that break the required natural balance to sustain diverse natural life. In such regard, humankind seems to constitute a plague that strives to transform the accessible universe to human image and need, disregarding the necessary ecological equilibrium to sustain the diversity of species and avoid their progressive elimination. Furthermore, current dominant conditions driven by dominant economic and political interests seem to feed on people's relatively passive, consumerist behaviour.

Could the human species and its cultural predominance reach further monopolistic levels? It would be feasible to anticipate the development of rigid social and behavioural patterns involving moral nature, social or political structures, cultural peculiarities, and abusive environmental decay. Perhaps, then, we cannot imagine a worse planet for humans, i.e., a uniform and rigid planetary community, although also more predictable—in this case, at the expense of individual and collective creativeness and freedom, which are fundamental values in the characterization of *Homo sapiens.*

The adaptability and plasticity of the species depend on several factors. Still, it is nourished by the genotypic and phenotypic variety and the possibility of expressing a range of physical and cultural profiles. Such spread expands the possible spectrum of the essential life adaptive condition of trial and error that characterizes the engine of natural evolution. The reduction in the degrees of cultural freedom would make humans and their communities intrinsically fragile in the long term. This is achieved by limiting their adaptive capacity and the emergence of suitable and successful phenotypes that continue to provide human civilization with original approaches and solutions to its problems and needs. The ancient criteria based on the success of the strongest warrior should be replaced by free interaction, the diversity of cultures and their profiles, and diverse contributions to humane forms of life.

Humans should have learned from their history of wars, destruction, and human slaughter, as displayed from primitive times to the present, and partially listed in Chapter 11. With time and the development of human skills, the warrior has replaced the hunter due to material dominance, religious beliefs, and/or ethnic disparities. Today, it is not only ethnic or belief differences that place humans at risk but also the validity of proposals of unique, exclusive, hegemonic thoughts—whether political, commercial, or religious.

Given the progress in biotechnological and scientific advances, humans should promote productive, proactive policies towards a viable, multicultural, equitable humanity, and avoid placing at-risk species involved in ecological equilibrium and a sustainable planet instead of transforming it into a barren field littered with remains from battles for temporary supremacy and unilateral, dominant socioeconomic development. The latter trend is continuously expressed by the increasing military spending by a group of military-leading countries, as published by *Statista* (2024):[7]

> The value of military spending globally has grown steadily in the past years and reached 2.44 trillion US dollars in 2023.

Though the dominance drive seems to be an engraved behaviour in animal species interactions, as discussed earlier (Colombo JA, 2022), among humans

> ... there should be an alternative for sociality based on solidarity, equal rights, and opportunity for collective development. Although there is a long way to utopia, our species should not abandon such a chimeric Horizon.

In conclusion, what significance does the concept of evolution have when applied to the human species in its social interaction and with the ecosystem? In a biological context, evolution has an interactive dominance and adaptive component related to survival. However, in this conceptual context, humans have overemphasized the emergence of new cognitive/technological dimensions and profit as defining features of human evolution. These late products of evolution affect the overall process and have impacted and derailed its evolutive path, altering environmental and species survival equilibrium. In an essential biological context, continuous environmental variations lead to ongoing adaptive changes in all organisms, adjusting survival capacities to new environmental conditions and demands. It is precisely this interactive process that appears menaced by human activity. Hence, it is crucial to critically evaluate whether, in the long term, human intervention in the ecosystem has a positive evolutionary—environmental interaction with species survival—or degrading effect. That is, that it does not act by eliminating species and as an environmental degradation agent, which seems to be the current condition. Though ecological changes and the transient survival and replacement of species have been natural, the anthropogenic effect responds to the direct actions of a single, prevailing macro-species.

To summarize, the basic concept is that biological species, particularly the human species, have undergone a constant dynamic of social dominance attempts with variable resistance, which has triggered among humans a race of increasing destructive power and asymmetrical development. This can undergo a general or localized annihilation encounter, placing humanity on the brink of compromising current environmental, social, behavioural, and survival needs and profiles.

## Notes

1 https://edition.cnn.com/2024/04/15/world/cicada-fungus-zombie-scn/index.html.
2 *Levels and trends in child mortality*. United Nations Inter-Agency Group for Child Mortality Estimation, Report 2023. Published March 2024.
3 https://www.gspublishing.com/content/research/en/reports/2023/03/27/d64e052b-0f6e-45d7-967b-d7be35fabd16.html.
4 https://www.gspublishing.com/content/research/en/reports/2023/03/27/d64e052b-0f6e-.45d7-967b-d7be35fabd16.html.
5 https://www.icanw.org/surge_2023_global_nuclear_weapons_spending.
6 https://www.sipri.org/yearbook/2023/07.
7 https://www.statista.com/statistics/262742/countries-with-the-highest-militaryspending/#:~:text=Global%20military%20spending,in%20the%20South%20China%20Sea.

## References


Beyrer C, Kamarulzaman A, Isbell M, Amon J, et al. 2024. Under threat: The International AIDS Society—*Lancet* Commission on Health and Human Rights. *The Lancet*. Commissions from the Lancet journals. https://doi.org/10.1016/SO140-6736(24)00302-7.

Boudry M, Hofhuis S. 2018. Parasites of the mind. Why cultural theorists need the meme's eye view. *Cogn Sys Res* 52: 155–167. https://doi.org/10.1016/j.cogsys.2018.06.010.

Briggs J, Kodnani D. 2023 *Global economics analyst. The potentially large effects of artificial intelligence on economic growth*. Goldman Sachs. https://www.gspublishing.com/content/research/en/reports/2023/03/27/d64e052b-0f6e-45d7-967b-d7be35fabd16.html.

Colombo E. 2013. *Historia del movimiento obrero revolucionario* (*History of the labor revolutionary movement*) (first Spanish edition). Libros de Anarres. ISBN: 09789871523184.

Colombo JA. 2007. *Acerca del desarrollo cerebral infantil* (*On brain child development*). In JA Colombo (editor), *Pobreza y desarrollo infantil* (*Poverty and child development*). Paidos Editorial.

Colombo JA. 2010. *Somos la especie equivocada?* (*Are humans the mistaken species?*). Eudeba. https://tinyurl.com/lsdozicg.

Colombo JA. 2015. *Los Homo sabios. Globalizados o segmentados?* (*The wise Homo. Globalized or segmented?*) Buenos Aires Books.

Colombo JA. 2019. *Our animal condition and social construction*. Nova Science Publishers. ISBN: 978-1-53615-357-6. eBook ISBN: 978-1-53615-3583.

Colombo JA. 2021a. Ancestral drives and the complex social behavior of human species. *Int. J Primatol Res*. https://www.researchgate.net/publication/349768145.

Colombo JA. 2021b. *The homo within the sapiens*. Nova Science Publishers. ISBN 978-1-53618-373-3. https://tinyurl.com/3wbdchlc.

Colombo JA. 2022. *Dominance behavior: An evolutive and comparative perspective.* Springer Nature.
Colombo JA. 2024a. *Evolution and the human-animal drive to conflict. A psychobiological perspective.* Routledge.
Colombo JA. 2024b. *Human Biome and Human Behaviour.* Routledge.
Crick F. 1995. *The astonishing hypothesis: The scientific search for the soul.* Simon & Schuster. ISBN 0-684-80158-2.
Dawkins R. 1976. *The selfish gene.* Oxford Landmark Science.
Hafer-Hahmann N. 2019. Experimental evolution of parasitic host manipulation. *Proc Biol Sci* 286(1895): 20182413. doi: 10.1098/rspb.2018.2413.
Hughes DP, Kibersat F. 2019. Parasite manipulation of host behavior. *Curr Biol* 29: R45–R47.
Poulin R. 1999. The functional importance of parasites in animal communities: Many roles at many levels? *Int J Parasitol* 29: 903–914. doi: 10.1016/s0020-7519(99)00045-4.

# INDEX

*Note*: Page numbers in *italics* indicate figures and in **bold** indicate tables on the corresponding pages.